Pragmatic Test-Driven Development in C# and .NET

Write loosely coupled, documented, and high-quality code with DDD using familiar tools and libraries

Adam Tibi

BIRMINGHAM—MUMBAI

Pragmatic Test-Driven Development in C# and .NET

Group Product Manager: Gebin George

Publishing Product Manager: Sathyanarayanan Ellapulli

Senior Editor: Nithya Sadanandan

Content Development Editor: Yashi Gupta

Technical Editor: Maran Fernandes

Copy Editor: Safis Editing

Project Coordinator: Deeksha Thakkar

Proofreader: Safis Editing

Indexer: Rekha Nair

Production Designer: Vijay Kamble

Developer Relations Marketing Executive: Sonakshi Bubbar

Business Development Executive: Kriti Sharma

First published: September 2022

Production reference: 1230922

Published by Packt Publishing Ltd.

Livery Place

35 Livery Street

Birmingham

B3 2PB, UK.

ISBN 978-1-80323-019-1

www.packt.com

I would like to thank my dear partner, Elvira, for keeping me on track, and for her support and encouragement throughout the book-writing journey. I wouldn't have done it without you. Also, thank you to my lovely sons, Robin and Charlie, for understanding why daddy was busy.

Also, I would like to thank the person who believed in me and enthused me, my teacher Dr. Jaber M. Jaber, and my favorite technical author of all time, Charles Petzold, who inspired me with his writing style and simplicity.

I would like to thank the editors, Nithya Sadanandan and Yashi Gupta, for their patience with me and for their remarks.

Finally, I would like to thank the reviewer, Ahmed Ilyas, for his comments and for his good grades for each chapter, which kept me going with passion.

Contributors

About the author

Adam Tibi is a London-based software consultant with over 22 years of experience in .NET, Python, the Microsoft stack, and Azure. He is experienced in mentoring teams, designing architecture, promoting agile and good software practices, and, of course, writing code. Adam has consulted for blue-chip firms including Shell, Lloyds Bank, Lloyd's of London, Willis Towers Watson, and for a mix of start-ups. As a consultant who has a heterogeneous portfolio of clients, he has gained a solid understanding of the TDD intricacies, which he has transferred into this book.

About the reviewers

Ahmed Ilyas has 18 years of professional experience in software development.

After leaving Microsoft, he ventured into setting up his consultancy company, offering the best possible solutions for a multitude of industries and providing real-world answers to those problems. He only uses the Microsoft stack to build these technologies and provide the best practices, patterns, and software to his client base to enable long-term stability and compliance in the ever-changing software industry, but also to help software developers around the globe to improve as well as push the limits of technology.

Three times awarded MVP in C# by Microsoft and with a great reputation earned, he has a large client base for his consultancy company, Sandler Software LLC, which includes clients from different industries. Clients have included him on their "approved contractors/consultants" list as a trusted vendor and partner and this further led to his joining Microsoft again.

He has been involved in reviewing books for Packt Publishing in the past and wishes to thank them for the great opportunity once again.

Table of Contents

3

Getting Started with Unit Testing   73

4

Real Unit Testing with Test Doubles   99

5

Test-Driven Development Explained   125

6

The FIRSTHAND Guidelines of TDD   151

Part 2: Building an Application with TDD

7

A Pragmatic View of Domain-Driven Design   181

8

Designing an Appointment Booking App   203

9

Building an Appointment Booking App with Entity Framework and Relational DB   219

10

Building an App with Repositories and Document DB   243

Part 3: Applying TDD to Your Projects

11

Implementing Continuous Integration with GitHub Actions   269

12

Dealing with Brownfield Projects   285

13

The Intricacies of Rolling Out TDD   303

Appendix 1: Commonly Used Libraries with Unit Tests   313

Appendix 2: Advanced Mocking Scenarios   327

Preface

As a consultant, I worked with many teams in multiple organizations. I've seen teams doing TDD and I've seen teams doing unit testing without TDD. I've also seen teams that thought they were doing unit testing but were doing integration testing and I've seen teams that are doing none! As an ordinary human being, I started forming a belief based on empirical evidence that TDD teams are the most successful, but this is not because they are using TDD! TDD results from having passion.

TDD is unit testing plus passion. Unit testing in some teams is imposed, therefore, the developers have to do it, but TDD is rarely imposed and it is up to the developers to enforce it on themselves. Needless to say, passionate developers produce quality projects and quality projects have more potential to succeed.

TDD is usually combined with some or all aspects of **domain-driven design** (**DDD**) architecture. So, I made sure I covered both TDD and DDD combined to be able to give realistic examples. I also wanted to reflect today's market that is divided between two database categories, relational and document DBs, so I took the liberty to include an example chapter for each and show the differences in unit-testing implementations with the objective of keeping the book pragmatic.

Don't be tricked by the book size, as the diagrams and the code snippets do inflate the book. I strived to stay away from old and impractical theories to shorten the book and stick to the point.

TDD and unit testing are in most modern job specifications, a requirement for interview test projects, and the subject of hot interview questions. If you wanted to know more about these topics and become a TDD developer, then you came to the right place.

There are many other good books on TDD and they are aimed at .NET developers as well, so *why this book?* In this book, I show real practical implementations by going to the DDD world, relational DBs, and document DBs. I demonstrate the decision tree of the mindset that practitioners use when doing TDD. I show the relationship between SOLID and TDD and I introduce a set of memorable best practices known as the FIRSTHAND guidelines of TDD.

My intention for writing this book is to have you as a confident TDD practitioner, or at least a unit-testing practitioner, and I hope I was able to deliver my intention.

Who this book is for

Test-driven development is the mainstream way of designing, documenting, and testing your application from day one. As a developer looking to climb the technical ladder to a more senior position, TDD and its related topics of unit testing, test doubles, and dependency injection are a must-learn.

This book is for mid- to senior-level .NET developers who are looking to utilize the potential of TDD to develop high-quality software. Basic knowledge of OOP and C# programming concepts is assumed, but no knowledge of TDD or unit testing is expected. As the book provides in-depth coverage of all the concepts of TDD and unit testing, and acts as an excellent guide for developers who want to build a TDD-based application from scratch or developers planning to introduce unit testing into their organization.

What this book covers

The book covers TDD and its .NET ecosystem of IDEs and libraries and goes through setting up the environment. The book starts by covering the topics that form the prerequisites for TDD, which are dependency injection, unit testing, and test doubles. Then, after covering TDD and its best practices, the book dives into building an application from scratch using domain-driven design as an architecture.

The book also covers the basics of building a continuous integration pipeline, dealing with legacy code that wasn't written with testability in mind, and finishing with ideas for rolling out TDD into your organization.

Chapter 1, *Writing Your First TDD Implementation*, doesn't have a long introduction or theory but rather dives directly into IDE selection and writing your first TDD implementation to get a taste of flavor of the content of the book.

Chapter 2, *Understanding Dependency Injection by Example*, revises advanced OOP principles needed to understand the concept of dependency injection and provides multiple examples.

Chapter 3, *Getting Started with Unit Testing*, offers a simple introduction to xUnit and unit-testing basics.

Chapter 4, *Real Unit Testing with Test Doubles*, goes through stubbing, mocking, and NSubstitute, and then discusses more testing categories.

Chapter 5, *Test-Driven Development Explained*, demonstrates how to write unit testing but in TDD style, and discusses the pros and cons.

Chapter 6, *The FIRSTHAND Guidelines of TDD*, details the best practices of unit testing and TDD.

Chapter 7, *A Pragmatic View of Domain-Driven Design*, introduces DDD, services, and repositories.

Chapter 8, *Designing an Appointment Booking App*, outlines the specification for a real-life app to be implemented later with a DDD architecture and the TDD style.

Chapter 9, *Building an Appointment Booking App with Entity Framework and Relational DB*, demonstrates an example of a TDD application using a relational DB backend.

Chapter 10, *Building an App with Repositories and Document DB*, demonstrates an example of a TDD application using a document DB and the repository pattern.

Chapter 11, *Implementing Continuous Integration with GitHub Actions*, shows how to use GitHub Actions to build a CI pipeline for the application in *Chapter 10*.

Chapter 12, *Dealing with Brownfield Projects*, outlines the thinking process when considering TDD and unit testing for legacy projects.

Chapter 13, *The Intricacies of Rolling Out TDD*, explains the train of thought when getting your organization to adopt TDD.

Appendix 1, *Commonly Used Libraries with Unit Tests*, shows some quick examples of MSTest, NUnit, Moq, Fluent Assertions, and Auto Fixture.

Appendix 2, *Advanced Mocking Scenarios*, demonstrates a more complex mocking scenario with NSubstitute.

To get the most out of this book

This book assumes you are familiar with C# syntax with at least one year's experience working with Visual Studio or a similar IDE environment. While the advanced concepts of OOP principles will be revised in this book, the book assumes you are familiar with the basics.

Software covered in the book	Operating system requirements
Visual Studio 2022	Windows or macOS
Fine Code Coverage	Windows
SQL Server	Windows, macOS (Docker), or Linux
Cosmos DB	Windows, macOS (Docker), or Linux (Docker)

Libraries and frameworks	Operating system requirements
.NET Core 6, C# 10	Windows, macOS, or Linux
xUnit	Windows, macOS, or Linux
NSubstitute	Windows, macOS, or Linux
Entity Framework	Windows, macOS, or Linux

To get the most out of the book, you need to have a C# IDE. This book uses Visual Studio 2022 Community Edition and presents alternatives at the beginning of *Chapter 1*.

If you are using the digital version of this book, we advise you to type the code yourself or access the code from the book's GitHub repository (a link is available in the next section). Doing so will help you avoid any potential errors related to the copying and pasting of code.

Download the example code files

You can download the example code files for this book from GitHub at https://github.com/
PacktPublishing/Pragmatic-Test-Driven-Development-in-C-Sharp-
and-.NET. If there's an update to the code, it will be updated in the GitHub repository.

We also have other code bundles from our rich catalog of books and videos available at https://
github.com/PacktPublishing/. Check them out!

Download the color images

We also provide a PDF file that has color images of the screenshots and diagrams used in this book.
You can download it here: https://packt.link/OzRlM.

Conventions used

There are a number of text conventions used throughout this book.

Code in text: Indicates code words in text, database table names, folder names, filenames, file
extensions, pathnames, dummy URLs, user input, and Twitter handles. Here is an example: "The
previous code fails this rule, as running UnitTest2 before UnitTest1 will fail the test."

A block of code is set as follows:

```
public class SampleTests
{
    private static int _staticField = 0;
    [Fact]
    public void UnitTest1()
    {
        _staticField += 1;
        Assert.Equal(1, _staticField);
    }
    [Fact]
    public void UnitTest2()
    {
        _staticField += 5;
        Assert.Equal(6, _staticField);
    }
}
```

When we wish to draw your attention to a particular part of a code block, the relevant lines or items are set in bold:

```
public class SampleTests
{
    private static int _staticField = 0;
    [Fact]
    public void UnitTest1()
    {
        _staticField += 1;
        Assert.Equal(1, _staticField);
    }
    [Fact]
    public void UnitTest2()
    {
        _staticField += 5;
        Assert.Equal(6, _staticField);
    }
}
```

Any command-line input or output is written as follows:

```
GET https://webapidomain/services
```

Bold: Indicates a new term, an important word, or words that you see onscreen. For instance, words in menus or dialog boxes appear in **bold**. Here is an example: "After installing the local emulator, you need to grab the connection string, which you can do by browsing to `https://localhost:8081/_explorer/index.html` and copying the connection string from the **Primary Connection String** field."

> **Tips or important notes**
> Appear like this.

Get in touch

Feedback from our readers is always welcome.

General feedback: If you have questions about any aspect of this book, email us at `customercare@packtpub.com` and mention the book title in the subject of your message.

Errata: Although we have taken every care to ensure the accuracy of our content, mistakes do happen. If you have found a mistake in this book, we would be grateful if you would report this to us. Please visit `www.packtpub.com/support/errata` and fill in the form.

Piracy: If you come across any illegal copies of our works in any form on the internet, we would be grateful if you would provide us with the location address or website name. Please contact us at `copyright@packt.com` with a link to the material.

If you are interested in becoming an author: If there is a topic that you have expertise in and you are interested in either writing or contributing to a book, please visit `authors.packtpub.com`.

Share Your Thoughts

Once you've read *Pragmatic Test-Driven Development in C# and .NET*, we'd love to hear your thoughts! Scan the QR code below to go straight to the Amazon review page for this book and share your feedback.

`https://packt.link/r/1803230193`

Your review is important to us and the tech community and will help us make sure we're delivering excellent quality content.

Part 1:
Getting Started and
the Basics of TDD

In this part, we will gradually introduce all the concepts that make up test-driven development – starting with dependency injection, going through test doubles, and ending with the TDD guidelines and best practices.

By the end of this part, you will have the necessary knowledge to contribute to an application using TDD. The following chapters are included in this part:

- *Chapter 1, Writing Your First TDD Implementation*
- *Chapter 2, Understanding Dependency Injection by Example*
- *Chapter 3, Getting Started with Unit Testing*
- *Chapter 4, Real Unit Testing with Test Doubles*
- *Chapter 5, Test-Driven Development Explained*
- *Chapter 6, The FIRSTHAND Guidelines of TDD*

Writing Your First TDD Implementation

I've always liked books that start with a quick end-to-end demo about the proposed subject before diving into the details. That gives me a sense of what I am going to learn. I wanted to share with you the same experience by beginning this book with a tiny application.

Here, we will simulate minimal business requirements, and while implementing them, we will touch on **unit testing** and **test-driven development** (**TDD**) concepts. Don't worry if a concept is not clear or requires further explanation, as this chapter purposely skims over topics to give you a flavor. By the end of the book, we will have covered all the concepts that were skimmed over.

Also, note that we will use the terms *unit testing* and *TDD* interchangeably with little distinction. The difference will be clearer by *Chapter 5, Test-Driven Development Explained*.

In this chapter, you will cover the following topics:

- Choosing your **integrated development environment** (**IDE**)
- Building a solution skeleton with unit testing
- Implementing requirements with TDD

By the end of the chapter, you will be comfortable writing basic unit tests using **xUnit** and have a fair understanding of what TDD is.

Technical requirements

The code for this chapter can be found at the following GitHub repository:

```
https://github.com/PacktPublishing/Pragmatic-Test-Driven-Development-
in-C-Sharp-and-.NET/tree/main/ch01
```

Choosing your IDE

From a TDD perspective, different IDEs will affect your productivity. TDD implementation can be boosted by IDEs that have rich code refactoring and code generation capabilities, and selecting the right one will reduce repetitive—and potentially boring—tasks.

In the following sections, I have presented three popular IDEs with C# support: **Visual Studio (VS)**, **VS Code**, and **JetBrains Rider**.

Microsoft VS

This chapter and the rest of the book will use *VS 2022 Community Edition*—this should also work with the *Professional* and *Enterprise* editions. Individual developers can use *VS Community Edition* for free to create their own free or paid applications. Organizations can also use it under some restrictions. For the full license and product details, visit `https://visualstudio.microsoft.com/vs/community/`.

If you have an earlier version of VS and do not want to upgrade, then you can have *VS 2022 Community Edition* installed side by side with previous versions.

Both the *Windows* and *Mac* editions of *VS 2022* have the required tools to build our code and run the tests. I have done all the projects, screenshots, and instructions in this book using the *Windows* edition. You can download VS from `https://visualstudio.microsoft.com/downloads/`.

When installing VS, you will need at least the **ASP.NET and web development** box selected to be able to follow along with the book, as illustrated in the following screenshot:

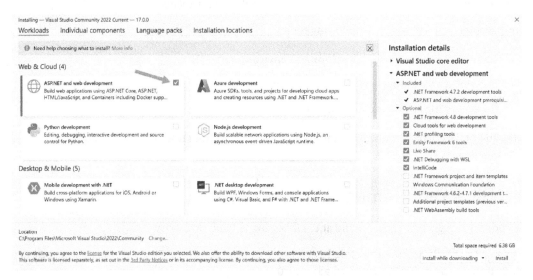

Figure 1.1 – VS installation dialog

If you have VS previously installed, you can check if **ASP.NET and web development** is already installed by following these steps:

1. Go to Windows **Settings | Apps | Apps & features**.

2. Search for Visual Studio under **App list**.

3. Select the vertical ellipsis (the three vertical dots).

4. Select **Modify**, as shown in the following screenshot:

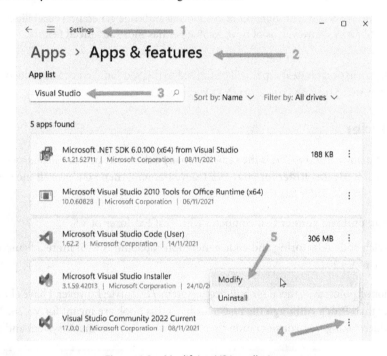

Figure 1.2 – Modifying VS installation

VS is big, as it contains plenty of components to install. Also, after installation, it is the slowest to load, in comparison with *Rider* and *VS Code*.

ReSharper

JetBrains ReSharper is a popular commercial plugin for VS. ReSharper adds multiple features to VS; however, from a TDD standpoint, we are interested in the following aspects:

*   **Refactoring**: ReSharper adds many refactoring features that come in handy when you reach the refactoring stage of TDD.

*   **Code generation**: Generating code with ReSharper is particularly useful when creating your unit tests first, then generating code after.

- **Unit testing**: ReSharper supercharges the unit testing tools in VS and has support for more unit testing frameworks.

ReSharper is a subscription-based product with a 30-day trial. I would recommend you to start first with VS without ReSharper, then add it later when you are familiar with the capabilities of VS so that you recognize the benefits of adding ReSharper.

> **Note**
> Each new release of VS adds additional code refactoring and code generation capabilities similar to those of ReSharper. However, as of now, ReSharper has more advanced features.

In this book, the discussion on ReSharper will be limited to this section. You can download ReSharper here: `https://www.jetbrains.com/resharper/`.

JetBrains Rider

JetBrains, the company behind *Rider*, is the same company behind the popular *ReSharper* VS plugin. If you have chosen **JetBrains Rider** for your .NET development, then you have all the features that are required in this book. Rider has the following:

- A powerful **unit test runner** that competes with **Test Explorer** of VS

- Feature-rich code refactoring and code generation capabilities with more advanced features than those of VS 2022

The aforementioned points are crucial for building a system *TDD-style*; however, I have chosen VS for this book rather than Rider. Although the instructions in this book are meant for VS 2022, they can be applied to Rider, taking into consideration that Rider has a different menu system and shortcuts.

> **Note**
> **VS .NET** (VS release with .NET support) was released in February 2002, while Rider is more recent and was released in August 2017; so, VS is more established between .NET developers. I have nominated VS for this book over Rider for this reason.

You can download Rider here: `https://www.jetbrains.com/rider/`.

VS Code

If you are a fan of VS Code, you will be pleased to know that Microsoft added native support for visual unit testing (which is essential for TDD) in July 2021, with the version 1.59 release.

VS Code is a lightweight IDE—it has good native refactoring options and a bunch of third-party refactoring plugins. The simplicity and elegance of VS Code attract many TDD practitioners, but the available C# features—especially those used in TDD—are not as advanced as those of VS or Rider.

I will be using *VS* in this book, but you can adapt the examples to VS Code where relevant. To download VS Code, you can visit `https://visualstudio.microsoft.com/downloads/`.

.NET and C# versions

VS 2022 comes with **.NET 6** and **C# 10** support. This is what we will be using for the purposes of this chapter and the rest of the book.

I initiated a small poll to gather some public opinion in my LinkedIn group—you can see the results here:

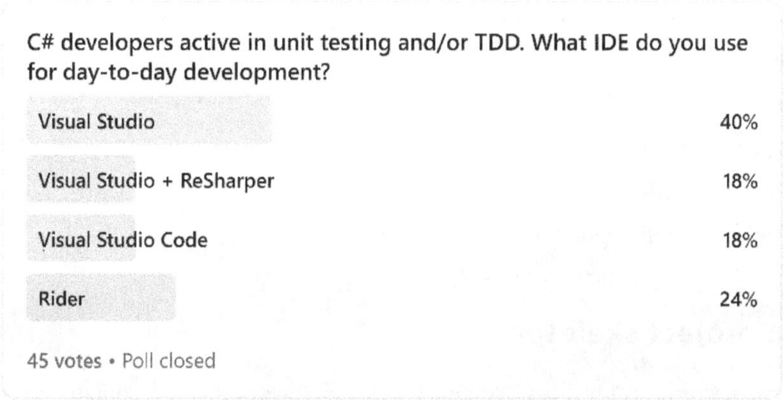

C# developers active in unit testing and/or TDD. What IDE do you use for day-to-day development?

Visual Studio	40%
Visual Studio + ReSharper	18%
Visual Studio Code	18%
Rider	24%

45 votes • Poll closed

Figure 1.3 – LinkedIn IDE poll results

As you can see, VS has the highest usage of 58%, with 18% who use the ReSharper plugin with VS, then Rider comes second at 24%, then in third place comes VS Code with 18%. However, given that this is only 45 votes, it is meant to give you an indication and would definitely not reflect the market.

Picking the right IDE is a debatable subject between developers. I know that every time I ask a developer practicing TDD about their chosen IDE, they would swear by how good their IDE is! In conclusion, use the IDE that makes you more productive.

Building a solution skeleton with unit testing

Now that we've got the technical requirements out of the way, it's time to build our first implementation. For this chapter, and to keep the focus on TDD concepts, let's start with simple business requirements.

Let's assume that you are a developer, working for a fictitious company called **Unicorn Quality Solutions Inc. (UQS)**, which produces high-quality software.

Requirements

The software teams in **UQS** follow an agile approach and describe business requirements in terms of a **user story**.

You are working on a math library that is packaged to be consumed by other developers. You can think of this as if you are building a feature within a **NuGet library** to be used by other applications. You've picked a user story to implement, as outlined here:

Story Title:

Integers Division

Story Description:

As a math library client, I want a method to divide two integers

Acceptance Criteria:

Supports an Int32 input and a decimal output

Supports high-precision return with no/minimal rounding

Supports dividing divisible and indivisible integers

Throws a DivideByZeroException when dividing by 0

Creating a project skeleton

You will need two C# projects for this story. One is the **class library** that will contain the production code and a second library for unit testing the class library.

Note

Class libraries enable you to modularize functionality that can be used by multiple applications. When compiled, they will generate **dynamic-link library** (DLL) files. A class library cannot run on its own, but it can run as part of an application.

If you have not worked with a class library before, for the purposes of this book, you can treat it like a console app or a web app.

Creating a class library project

We are going to create the same projects set up in two ways—via the **graphical user interface** (GUI) and via the .NET **command-line interface** (CLI). Choose what you prefer or what you are familiar with.

Via the GUI

To create a class library, run VS and then follow these steps:

1. From the menu, select **File | New | Project**.
2. Look for Class Library (C#).
3. Select the rectangle containing **Class Library (C#) |** hit **Next**. The **Add a new project** dialog will display, as follows:

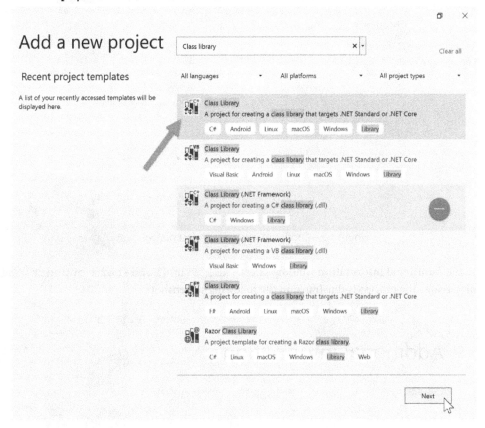

Figure 1.4 – Finding the Class Library (C#) project template

Important Note

Make sure you can see the **C#** tag in the box and do *NOT* select the **Class Library (.NET Framework)** item. We are using .NET (not classical .NET Framework).

4. In the **Configure your new project** dialog, type `Uqs.Arithmetic` in the **Project name** field and `UqsMathLib` in the **Solution name** field, then hit **Next**. The process is illustrated in the following screenshot:

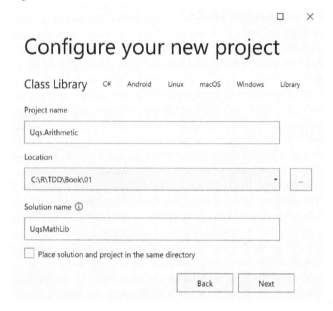

Figure 1.5 – Configure your new project dialog

5. In the **Additional information** window, select `.NET 6.0 (Long-term support)` and hit **Create**. The process is illustrated in the following screenshot:

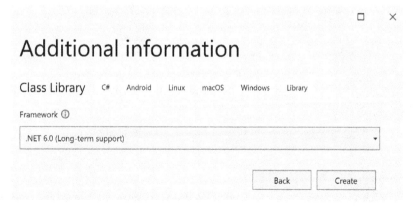

Figure 1.6 – Additional information

We now have a class library project within the solution, using the *VS GUI*.

Via the CLI

If you prefer to create your project via the CLI, here are the required commands:

1. Create a directory called UqsMathLib (md UqsMathLib).

2. Navigate to this directory via your terminal (cd UqsMathLib), as illustrated in the following screenshot:

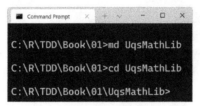

Figure 1.7 – Command Prompt showing the commands

3. Create a solution file (.sln) that will be generated with the same name as the directory—that is, UqsMathLib.sln—by running the following command:

```
dotnet new sln
```

4. Create a new class library called Uqs.Arithmetic in a directory with the same name and use *.NET 6.0*. Here's the code you need to execute:

```
dotnet new classlib -o Uqs.Arithmetic -f net6.0
```

5. Add the newly created project to the solution file by running the following command:

```
dotnet sln add Uqs.Arithmetic
```

We now have a class library project within the solution, using the *CLI*.

Creating a unit testing project

Currently, we have a solution with one class library project. Next, we want to add the unit test library to our solution. For this, we will use **xUnit Test Project**.

xUnit.net is a free, open source, unit testing tool for .NET. It is licensed under Apache 2. VS natively supports adding and running xUnit projects, so no special tool or plugin is needed to use xUnit.

We will be going into more details about xUnit in *Chapter 3, Getting Started with Unit Testing*.

We will follow a common convention for naming unit test projects: [ProjectName].Tests.Unit. Thus, our project will be called Uqs.Arithmetic.Tests.Unit.

We will create a unit test project in two ways, so you can pick whatever suits you best.

Via the GUI

To create a unit testing project, go to **Solution Explorer** in VS, then follow these steps:

1. Right-click on the solution file (UqsMathLib).
2. Go to **Add | New Project…**, as illustrated in the following screenshot:

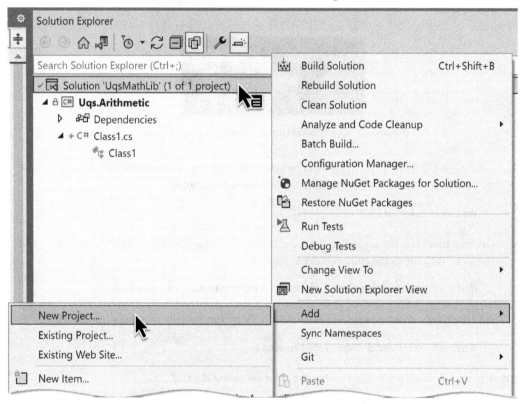

Figure 1.8 – Creating a new project in a solution

3. Look for **xUnit Test Project** | hit **Next**.
4. Set the **Project name** value as Uqs.Arithmetic.Tests.Unit.
5. Hit **Next** | select **.NET 6.0** | hit **Create**.

You have created a project via the VS GUI, but we still need to set the unit test project to have a reference to the class library. To do so, follow these steps:

1. In VS Solution Explorer, right-click on **Dependencies** in Uqs.Arithmetic.Tests.Unit.

2. Select **Add Project Reference…**.

3. Tick Uqs.Arithmetic and hit **OK**.

We now have our solution fully constructed via the VS GUI. You may choose to do the same GUI steps in the CLI instead. In the next section, we will do exactly that.

Via the CLI

Currently, we have a solution with one class library project. Now, we want to add the unit test library to our solution.

Create a new xUnit project called Uqs.Arithmetic.Tests.Unit in a directory with the same name and use .NET 6.0. Here's the code you need to execute:

```
dotnet new xunit -o Uqs.Arithmetic.Tests.Unit -f net6.0
```

Add the newly created project to the solution file by running the following command:

```
dotnet sln add Uqs.Arithmetic.Tests.Unit
```

We now have two projects in our solution. As the unit test project will be testing the class library, the project should have a reference to the class library.

You have created a project via the CLI, but we still need to set the unit test project to have a reference to the class library. To do so, add a project reference from Uqs.Arithmetic.Tests.Unit to Uqs.Arithmetic, like so:

```
dotnet add Uqs.Arithmetic.Tests.Unit reference
    Uqs.Arithmetic
```

We now have our solution fully constructed via the CLI.

Final solution

Whichever method you've used to create the solution—either the VS GUI or the CLI—you should now have the same files created. Now, you can open the solution in VS, and you'll see this:

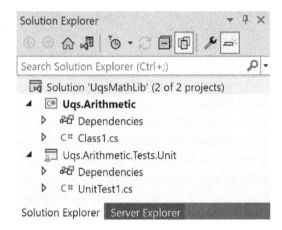

Figure 1.9 – The final created solution structure

To start from a clean slate, delete `Class1.cs` as we won't be using it—it was added automatically by the template.

The logical structure of our two projects looks like this:

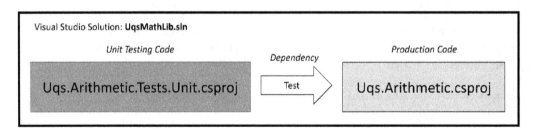

Figure 1.10 – Projects' logical structure

What we've created so far are two projects: one that will be shipped to production at some stage (`Uqs.Arithmetic`) and one to test this project (`Uqs.Arithmetic.Tests.Unit`). The solution file links the two projects together.

Now that we have finished the less fun part of building the project skeleton and setting the dependencies, we can now start the more fun one, which is directly related to unit testing.

Familiarizing yourself with built-in test tools

We have reached the stage where we need to look into how to discover and execute tests, and to do that, we need to understand which tools are available to us.

We have code that is already generated by the xUnit template—look at the code inside UnitTest1. cs, as displayed here:

```
using Xunit;
namespace Uqs.Arithmetic.Tests.Unit;


public class UnitTest1
{
    [Fact]
    public void Test1()
    {
    }
}
```

This is a normal C# class. Fact is an attribute from xUnit. It simply tells any xUnit-compatible tool that the method decorated with Fact is a **unit test method**. xUnit-compatible tools such as **Test Explorer** and **.NET CLI Test Command** should be able to find this method in your solution and run it.

Following the trend of the previous sections, we will utilize the available test tools in two ways—via the VS GUI and via the CLI.

Via the GUI

VS comes with a GUI as a test runner to discover and execute tests—it is called **Test Explorer**. To see how a test runner would discover test methods, from the menu, go to **Test | Test Explorer**. You will see the following screen:

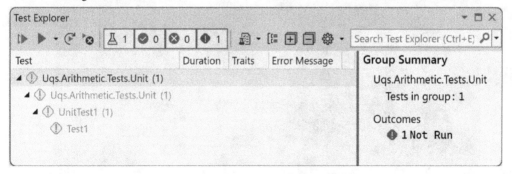

Figure 1.11 – Test Explorer showing unexecuted tests

As you can see, it detected all the tests in our solution, showing the tests in a *Project Name > Namespace > Class > Method* hierarchy. Also, you can see that the test hierarchy is grayed out and has an exclamation mark. This is a sign that the tests were never run. You can click on the upper-left **Run** button (shortcut keys: *Ctrl + R, T*. This is pressing and holding *Ctrl* then pressing *R* and quickly switching from *R* to *T*) and run this test. This will build your project and will execute the code within the methods decorated with `Fact`. The results are shown here:

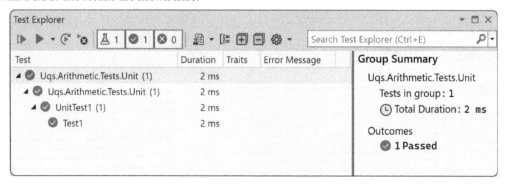

Figure 1.12 – Test Explorer showing executed test results

Don't expect anything fancy as we have an empty shell, but at least the tests will turn *green* and you will know that your setup is working. You can similarly discover and execute tests using the CLI.

Via the CLI

You can also execute the same test by using Command Prompt, going to the solution directory, and executing the following command:

```
dotnet test
```

This is what you are going to get:

```
C:\R\TDD\Book\01\UqsMathLib>dotnet test
  Determining projects to restore...
  All projects are up-to-date for restore.
  Uqs.Arithmetic -> C:\R\TDD\Book\01\UqsMathLib\Uqs.Arithmetic\bin\Debug\net6.0\Uqs.Arithmetic.dll
  Uqs.Arithmetic.Tests.Unit -> C:\R\TDD\Book\01\UqsMathLib\Uqs.Arithmetic.Tests.Unit\bin\Debug\net6.0\
Uqs.Arithmetic.Tests.Unit.dll
Test run for C:\R\TDD\Book\01\UqsMathLib\Uqs.Arithmetic.Tests.Unit\bin\Debug\net6.0\Uqs.Arithmetic.Tes
ts.Unit.dll (.NETCoreApp,Version=v6.0)
Microsoft (R) Test Execution Command Line Tool Version 17.0.0
Copyright (c) Microsoft Corporation.  All rights reserved.

Starting test execution, please wait...
A total of 1 test files matched the specified pattern.

Passed!  - Failed:     0, Passed:     1, Skipped:     0, Total:     1, Duration: < 1 ms - Uqs.Arithmet
ic.Tests.Unit.dll (net6.0)

C:\R\TDD\Book\01\UqsMathLib>
```

Figure 1.13 – .NET Test command discovering and executing tests

Running commands such as this will come in handy later on when we want to automate test running.

Implementing requirements with TDD

Before writing any code, it makes sense that we understand some terminologies and conventions to tune our brain on unit test-related keywords. So, we will briefly touch on **system under test (SUT)**, **red/green** tests, and **Arrange-Act-Assert (AAA)**. More details on these terminologies will follow in later chapters, but for now, we will cover the minimum to get a few tests running.

While we are learning about terminology and conventions, we will ease into our implementation. One thing that you might find new or unordinary is writing a unit test first, then writing production code later. This is one main aspect of TDD, and you will first experience it in this section.

SUT

We refer to the code that you usually write to build a product as **production code**. Typical **object-oriented (OO)** production code looks like this:

```
public class ClassName
{
    public Type MethodName (...)
    {
        // Code that does something useful
    }
    // more code
}
```

When we test this code, the unit test will call `MethodName` and assess the behavior of this method. When `MethodName` is executed, it may call other parts of the class and may use/call other classes. The code executed by `MethodName` is called SUT or **code under test (CUT)**. However, the term *SUT* is used more often.

The SUT will have an entry point that will be executed by the unit tests. The entry point is usually the method that we are calling from the unit tests. The following screenshot should clarify the idea of a SUT and a SUT entry point:

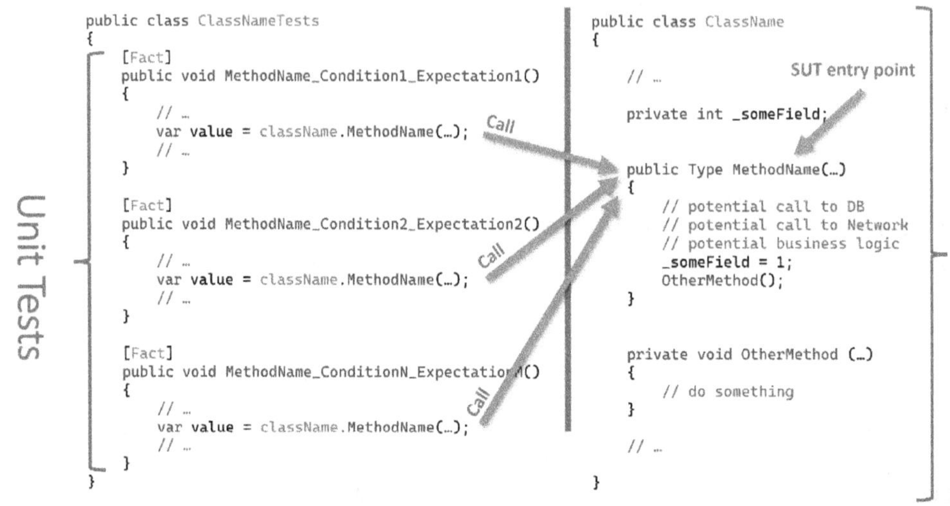

Figure 1.14 – Unit tests operating on a SUT

In the previous screenshot, you can see multiple unit tests calling the same SUT entry point. A detailed discussion of SUT is available in *Chapter 3, Getting Started with Unit Testing*.

Testing class

A typical unit testing class uses the same names from the SUT, by convention. This is what a typical unit testing class looks like:

```
public class ClassNameTests
{

    [Fact]
    public void MethodName_Condition1_Expectation1()
    {
        // Unit Testing Code that will call MethodName
    }
    // Other tests...
    [Fact]
    public void MethodName_ConditionN_ExpectationN()
    {
```

```
                    // Unit Testing Code that will call MethodName
        }
        ...
    }
```

Notice that the `ClassName` and `MethodName` methods in the two previous snippets are not a coincidence. We want them to be the same, again, by convention. To start forming our test class, we need to design the class name and the method name.

Class name

From the requirements, we will need a class that will contain all our division methods, so let's simply call the class `Division`; and if we were to create a unit test class to test the `Division` class, our unit test name would be called `DivisionTests`. Next, we will rename the `UnitTest1` class `DivisionTests` and rename the file as well so that it appears as `DivisionTests.cs`.

> **Tip**
> You can set your text cursor anywhere within the class name in the source code (in the previous case, it was `UnitTest1`) and hit *Ctrl + R, R* (hold *Ctrl* then press *R* quickly twice). Type the new name `DivisionTests` and hit *Enter*. This will also rename the file if the **Rename symbol's file** checkbox is ticked.

Method name

Luckily, the requirements are simple, so our method name will simply be `Divide`. `Divide` will be accepting two integer (`int32`) arguments, per the requirements, and returns a `decimal` value. We will go ahead and refactor our existing unit test from `Test1` to `Divide_Condition1_Expectation1`.

> **Note**
> **Arithmetic terminology-naming reminder**: If we have 10 / 5 = 2, then 10 is the dividend, 5 is the divisor, and 2 is the quotient.

Conditions and expectations

When we test, we are setting a condition and defining what we expect when this condition is met. We start with the *core case*, also known as the *positive path* or the *happy path*. We finish all the positive paths first before going to other cases. Our mission in unit tests boils down to determining the condition and its expectation and having a unit test for every combination.

To show the relationship between the method we are testing (the method in our SUT) and the associated condition and expectation, we will employ a well-used convention, as illustrated in the following code snippet:

```
[Fact]
public void MethodName_Condition_Expectation()
{
    ...
```

Here are random examples of unit test method names to familiarize you with the previous convention:

- `SaveUserDetails_MissingEmailAddress_EmailIsMissing`
- `ValidateUserCredentials_HashedPasswordDoesntMatch_False`
- `GetUserById_IdDoesntExist_UserNotFoundException`

We will see more examples while designing our unit tests.

The core requirement is dividing two integers. The straightforward and simplest implementation is dividing two divisible integers and getting back a whole number. Our condition is *divisible integers* and we expect a *whole number*. Now, we should update the signature of our unit test to `Divide_DivisibleIntegers_WholeNumber` and write the body of the test method, as follows:

```
[Fact]
public void Divide_DivisibleIntegers_WholeNumber()
{
    int dividend = 10;
    int divisor = 5;
    decimal expectedQuotient = 2;

    decimal actualQuotient = Division.Divide(dividend,
        divisor);

    Assert.Equal(expectedQuotient, actualQuotient);
}
```

This code doesn't compile as the `Division` class doesn't exist at this stage, and we know that already as we have a squiggly line under `Division`. This is one of the rare occasions where not being able to compile due to a missing class is good. This indicates that our *test has failed*, which is also good!

While it does look silly that the test has failed because the code doesn't compile as the `Division` SUT class is missing, this means that there is no SUT code yet. In *Chapter 5, Test-Driven Development Explained*, we will understand the reason behind considering the no-compilation case.

`Assert` is a class from the xUnit library. The `Equal` static method has many overloads, one of which we are using here:

```
public static void Equal<T>(T expected, T actual)
```

When run, this method will flag to the xUnit framework if what we expect and what we've actually got are equal. When we run this test, if the result of this assertion is `true`, then the test has passed.

Red/green

Failure is what we were looking for. In later chapters, we will discuss why. For now, it is sufficient to know that we need to start with a failed build (compilation) or failed test (failed assertion), then change that to a passed one. The fail/pass is also known as the **red/green refactor technique**, which mimics the idea of *bad/good* and *stop/go*.

We need to add the `Division` class and the `Divide` method and write the minimal code to make the test pass. Create a new file called `Division.cs` in the `Uqs.Arithmetic` project, like this:

```
namespace Uqs.Arithmetic;

public class Division
{
    public static decimal Divide(int dividend, int divisor)
    {
        decimal quotient = dividend / divisor;
        return quotient;
    }
}
```

> **Tip**
>
> You can create a class by placing the text cursor anywhere within the class name (in the previous case, it was `Division`) and hitting *Ctrl + .* (hold down the *Ctrl* key and then press .). Select **Generate new type…**, then from the **Project** dropdown, select `Uqs.Arithmetic`, and then hit **OK**. Then, to generate the method, place your text cursor on `Divide` and hit *Ctrl + .*, select **Generate method 'Division.Divide'**, and then you get the method shell in `Division` ready for your code.

It is important to remember that *dividing two integers in C# will return an integer*. I have seen senior developers fail to remember this, which led to bad consequences. In the code that we implemented, we have only covered the integers division that will yield a whole quotient. This should satisfy our test.

We are now ready to run our test with Test Explorer, so hit *Ctrl + R, A*, which will build your projects, then run all the tests (currently one test). You'll notice that Test Explorer indicates green, and there is a green bullet with a tick mark between the test name and the `Fact` attribute. When clicked, it will show you some testing-related options, as illustrated in the following screenshot:

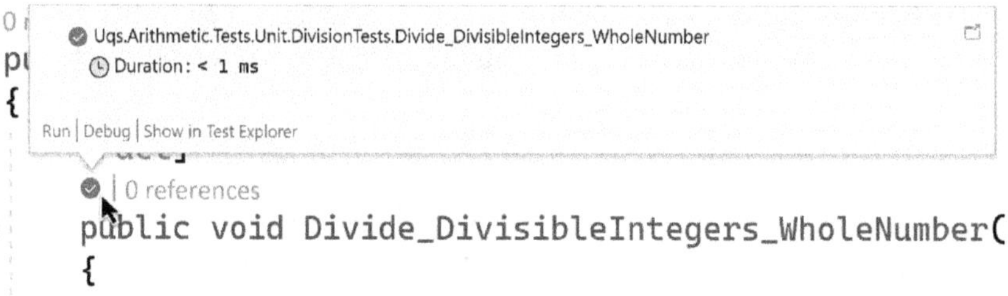

Figure 1.15 – VS unit testing balloon

For the sake of completion, the full concept name is **red/green/refactor**, but we won't be explaining the **refactor** bit here and will leave this for *Chapter 5, Test-Driven Development Explained*.

The AAA pattern

Unit testing practitioners noticed that test code format falls into a certain structure pattern. First, we declare some variables and do some preparations. This stage is called **Arrange**.

The second stage is when we invoke the SUT. In the previous test, it was the line on which we called the `Divide` method. This stage is called **Act**.

The third stage is where we validate our assumption—this is where we have the `Assert` class being used. This stage is, not surprisingly, called **Assert**.

Developers usually divide each unit test with comments to denote these three stages, so if we apply this to our previous unit test, the method would look like this:

```
[Fact]
public void Divide_DivisibleIntegers_WholeNumber()
{
    // Arrange
    int dividend = 10;
    int divisor = 5;
```

```
    decimal expectedQuotient = 2;

    // Act
    decimal actualQuotient = Division.Divide(dividend,
        divisor);

    // Assert
    Assert.Equal(expectedQuotient, actualQuotient);
}
```

You can learn more about the **AAA** pattern in *Chapter 3, Getting Started with Unit Testing.*

More tests

We haven't finished implementing the requirements. We need to add them iteratively, by adding a new test, checking that it fails, implementing it, then making it pass, and then repeating it!

We are going to add a few more tests in the next sections to cover all the requirements, and we are also going to add some other tests to increase the quality.

Dividing two indivisible numbers

We need to cover a case where two numbers are not divisible, so we add another unit testing method under the first one, like so:

```
[Fact]
public void Divide_IndivisibleIntegers_DecimalNumber()
{
    // Arrange
    int dividend = 10;
    int divisor = 4;
    decimal expectedQuotient = 2.5m;

    ...

}
```

This unit test method is similar to the previous one, but the name of the method has changed to reflect the new condition and expectation. Also, the numbers have changed to fit the new condition and expectation.

Run the test by employing any of the following methods:

- Clicking the blue bullet that appears below Fact, then clicking **Run**
- Opening **Test** | **Test Explorer**, selecting the new test name code, and clicking the **Run** button
- Pressing *Ctrl + R, A*, which will run all tests

You will notice that the test will fail—this is good! We have not implemented the division that will yield a decimal yet. We can go ahead and do it now, as follows:

```
decimal quotient = (decimal)dividend / divisor;
```

> **Note**
>
> Dividing two integers in C# will return an integer, but dividing a decimal by an integer returns a decimal, therefore you almost always have to cast the dividend or the divisor—or both—to a decimal.

Run the test again, and this time it should pass.

Division-by-zero test

Yes—bad things happen when you divide by zero. Let's check whether our code can handle this, as follows:

```
[Fact]
public void Divide_ZeroDivisor_DivideByZeroException()
{
    // Arrange
    int dividend = 10;
    int divisor = 0;

    // Act
    Exception e = Record.Exception(() =>
        Division.Divide(dividend, divisor));

    // Assert
    Assert.IsType<DivideByZeroException>(e);
}
```

The `Record` class is another member of the xUnit framework. The `Exception` method records whether the SUT has raised any `Exception` object and returns `null` if there is none. This is the method's signature:

```
public static Exception Exception(Func<object> testCode)
```

`IsType` is a method that compares the class type between the angle brackets to the class type of the object that we passed as an argument, as illustrated in the following code snippet:

```
public static T IsType<T>(object @object)
```

When you run this test, it will pass! My first impression would be one of suspicion. The problem is that when it passes without writing explicit code, we don't know yet whether this is a true or a coincidental pass—a false positive. There are many ways to validate whether this pass is incidental; the quickest way—for now—is to debug the code of `Divide_ZeroDivisor_DivideByZeroException`.

Click the **Test Bullet**, and then click the **Debug** link, as illustrated in the following screenshot:

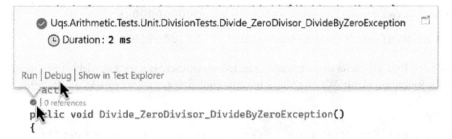

Figure 1.16 – The Debug option in the unit testing balloon

You will hit the exception directly, as illustrated in the following screenshot:

```
3 references | ● 3/3 passing
public static decimal Divide(int dividend, int divisor)
{
    decimal quotient = (decimal)dividend / divisor;  ⊗
    return quotient
}
```

Exception User-Unhandled   ▶ ⊡ ✕

System.DivideByZeroException: 'Attempted to divide by zero.'

View Details │ Copy Details │ Start Live Share session…
▷ Exception Settings

Figure 1.17 – Exception dialog

You'll notice that the exception is happening at the right place at the division line, so this is what we actually wanted. While this method violated our initial attempt of red/green, having a pass immediately is still a genuine case that you would encounter in day-to-day coding.

Testing extremes

The story did not mention testing the extremes, but as a developer, you know that most software bugs come from **edge cases**.

You want to build more confidence in your existing code, and you want to make sure that it can handle extremes well, as you'd expect it to.

The extreme values of an `int` data type can be obtained by these two constant fields of `int`:

- `int.MaxValue` = 2147483647 = $2^{31} - 1 - 1$
- `int.MinValue` = -2147483648 = -2^{31}

What we need to do is to test the following cases (note that we will only test for 12 decimal digits):

- `int.MaxValue / int.MinValue = -0.999999999534`
- `(-int.MaxValue) / int.MinValue = 0.999999999534`
- `int.MinValue / int.MaxValue = -1.000000000466`
- `int.MinValue / (-int.MaxValue) = 1.000000000466`

So, we will need four unit tests to cover each case. However, there is a trick available in most unit test frameworks, including xUnit. We don't have to write four unit tests—we can do this instead:

```
[Theory]
[InlineData( int.MaxValue,  int.MinValue, -0.999999999534)]
[InlineData(-int.MaxValue,  int.MinValue,  0.999999999534)]
[InlineData( int.MinValue,  int.MaxValue, -1.000000000466)]
[InlineData( int.MinValue, -int.MaxValue,  1.000000000466)]
public void Divide_ExtremeInput_CorrectCalculation(
    int dividend, int divisor, decimal expectedQuotient)
{
    // Arrange

    // Act
    decimal actualQuotient = Division.Divide(dividend,
        divisor);

    // Assert
    Assert.Equal(expectedQuotient, actualQuotient, 12);
}
```

Notice that now we have `Theory` rather than `Fact`. This is xUnit's way of declaring that the unit test method is parametrized. Also, notice that we have four `InlineData` attributes; as you will have already figured out, each one of them corresponds to a test case.

Our unit test method and the `InlineData` attributes have three parameters. When running the unit tests, each parameter will map to the corresponding unit test method's parameter in the same order. The following screenshot shows how each parameter in the `InlineData` attribute corresponds to a parameter in the `Divide_ExtremeInput_CorrectCalculation` method:

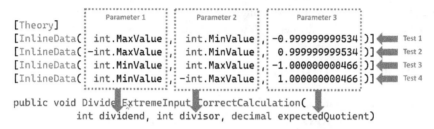

Figure 1.18 – InlineData parameters are mapped to the decorated method parameters

For assertion, we are using an overload of the `Equal` method that supports decimal **precision**, as illustrated in the following code snippet:

```
static void Equal(decimal expected, decimal actual,
    int precision)
```

Run the tests, and you'll notice that Test Explorer treats the four attributes as separate tests, as depicted in the following screenshot:

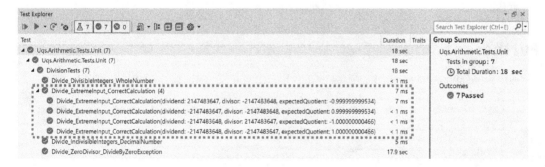

Figure 1.19 – VS Test Explorer showing grouped tests

> **Even More Tests**
>
> For brevity, and given that this chapter is a limited introduction, we didn't explore all possible testing scenarios—take, for example, `int.MaxValue/int.MaxValue`, `int.MinValue/int.MinValue`, `0/number`, and `0/0`.

The limits of the required tests are going to be discussed in later chapters, along with their pros and cons.

Writing tests before writing the code is not to every developer's taste and might look unintuitive at the beginning, but you have a complete book to make you decide for yourself. In *Chapter 5, Test-Driven Development Explained*, you will dig deeper into implementation and best practices.

Summary

While this chapter is meant for a quick implementation, I trust you did have a taste of what TDD is and picked up some skills, such as *xUnit*, *Test Explorer*, *test first*, *red/green*, and a few conventions.

For a start, we have picked easy examples—of course—so, we have got no **dependency injection (DI)**, nor mocking or anything fancy, because the stimulating stuff is coming next. So, I hope this chapter has made you excited about the rest of the book.

If you are like me when I first encountered TDD, you might be wondering the following: *Why test first? Isn't this too much unit testing code? Is unit testing effective? What is the difference between unit testing and TDD? How many tests should I write?* You may have other questions too—these will be answered gradually while you progress through the book, and I promise I will make the answers as clear as possible.

In the next chapter, we will touch on a design pattern called DI, which is an essential requirement for working with TDD.

Further reading

To learn more about the topics discussed in the chapter, you can refer to the following links:

- Class libraries: `https://docs.microsoft.com/en-us/dotnet/standard/class-libraries`
- xUnit: `https://xunit.net/`

2

Understanding Dependency Injection by Example

Dependency injection (DI) is a software design pattern that exists in every modern architecture. However, you may wonder how this pattern found its way into the second chapter of a **test-driven development (TDD)**-focused book.

DI is a pattern that has several benefits that we are going to discover throughout the book, though the core benefit is that *DI opens an application for unit testing*. We cannot exercise unit testing without a solid understanding of this pattern, and if we cannot unit test, by virtue, we cannot practice TDD. Considering this, DI understanding forms the foundation of *Section 1, Getting Started and Basics*, and *Part 2, Building an Application with TDD*, which explains the early introduction.

We will build an application and then modify it to support DI while learning the concepts, but the ideas in this chapter will be repeated and exercised throughout this book.

In this chapter, you will be exploring these topics:

- The **weather forecaster application (WFA)**
- Understanding dependency
- Introducing DI
- Using DI containers

By the end of this chapter, the application will be unit test-ready by having the necessary DI changes implemented. You will have a fair understanding of dependency and will have gained confidence in refactoring code to support DI. You will have also covered half the way to write your first proper unit test.

Technical requirements

The code for this chapter can be found at the following GitHub repository:

`https://github.com/PacktPublishing/Pragmatic-Test-Driven-Development-in-C-Sharp-and-.NET/tree/main/ch02`

There, you will find four directories. Each one will be a snapshot of our progress.

The WFA

Throughout this chapter, we will be using an **ASP.NET Web API** application in our learning process. We will be refactoring all the code in this application to enable DI. Then, in *Chapter 3, Getting Started with Unit Testing*, we will apply unit tests on the refactored application.

When a new ASP.NET Web API application is created, it comes with a sample random weather forecaster. The application in this chapter will build on top of the original weather sample and will add a real weather forecasting capability to the existing random one. We will creatively call our application the WFA.

The first step is going to be creating a WFA application and making sure it is running.

Creating a sample weather forecaster

To create a sample application, navigate your console to the directory where you want to create this application and execute the following commands:

```
md UqsWeather
cd UqsWeather
dotnet new sln
dotnet new webapi -o Uqs.Weather -f net6.0
dotnet sln add Uqs.Weather
```

The preceding code will create a **Visual Studio** (**VS**) solution called `UqsWeather` and will add an ASP.NET Web API project to it. This will produce a similar output to this console window:

```
C:\R\TDD\Book\ch02>md UqsWeather

C:\R\TDD\Book\ch02>cd UqsWeather

C:\R\TDD\Book\ch02\UqsWeather>dotnet new sln
The template "Solution File" was created successfully.

C:\R\TDD\Book\ch02\UqsWeather>dotnet new webapi -o Uqs.Weather -f net6.0
The template "ASP.NET Core Web API" was created successfully.

Processing post-creation actions...
Running 'dotnet restore' on C:\R\TDD\Book\ch02\UqsWeather\Uqs.Weather\Uqs.Weather.csproj...
  Determining projects to restore...
  Restored C:\R\TDD\Book\ch02\UqsWeather\Uqs.Weather\Uqs.Weather.csproj (in 322 ms).
Restore succeeded.

C:\R\TDD\Book\ch02\UqsWeather>dotnet sln add Uqs.Weather
Project `Uqs.Weather\Uqs.Weather.csproj` added to the solution.

C:\R\TDD\Book\ch02\UqsWeather>
```

Figure 2.1 – The output of creating a weather application via the command-line interface (CLI)

To check what we have created, go to the directory and open the solution using VS, and you will see the following:

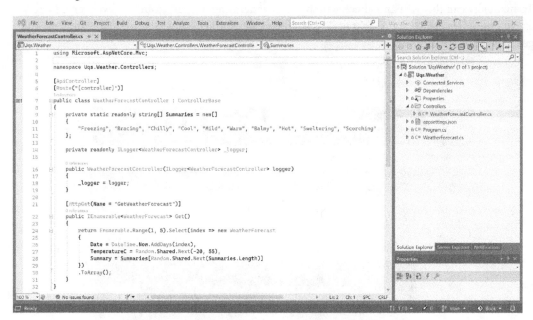

Figure 2.2 – Newly created project opened in VS

What is interesting here is the sample files that were generated automatically: `WeatherForecast Controller.cs` and `WeatherForecast.cs`.

This is the default template; we haven't done any modification yet. It would make sense to check whether, so far, the application is loading properly. You can run the application and it will launch your default browser with the Swagger UI interface. We can see the only available GET **application programming interface (API)**, `WeatherForecast`, as illustrated in the following screenshot:

Figure 2.3 – Swagger UI showing the available GET API

To manually call this API and check whether it is generating output, from the Swagger UI page, expand the down arrow on the right of **/WeatherForecast**. Hit **Try it out**. Then, hit **Execute**. You will get a response like this:

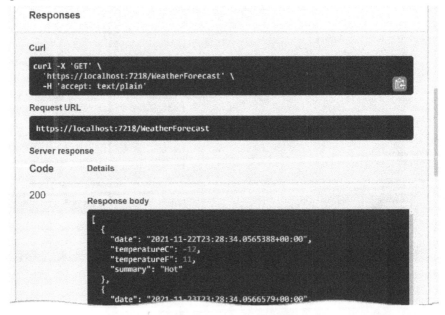

Figure 2.4 – Swagger API call response

You can find this sample under the GitHub chapter directory, in a directory called `01-UqsWeather`. Now, it's time to make the application a bit realistic by adding a real forecasting feature.

Adding a real weather forecaster

The template app has a sample random weather generator. I decided to give the app a spin by adding a real weather forecast as well. For this, I am going to use a weather service called *OpenWeather*. *OpenWeather* provides a free RESTful API weather service (where **REST** stands for **REpresentational State Transfer**) and will act as a more realistic example.

I have also created a public NuGet package to serve the chapter and act as a client for the *OpenWeather* RESTful APIs. So, rather than dealing with the REST API calls, you call a C# method, and it does the RESTful API calls in the background. In the following sections, we will obtain an API key and write the `GetReal` API.

Getting an API key

To be able to run the application from the companion source code or to create one yourself, you need an API key. You can sign up at `https://openweathermap.org` and then obtain an API key. After signing up, you can generate a key by going to **My API keys** and hitting **Generate**, similar to this:

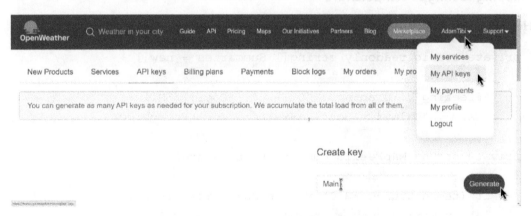

Figure 2.5 – Generating an API key

Once you have obtained the key, save it in your `appsettings.json` file, like this:

```
{
    "OpenWeather": {
      "Key": "yourapikeygoeshere"
    },
```

```
    "Logging": {
    ...
```

The API key is done. Let's get a client library to access the API.

Getting the client NuGet package

There are many OpenWeather API client libraries; however, I chose to create one to specifically fit the requirements of this chapter. The code of the package and how it is tested are discussed in *Appendix 2, Advanced Mocking Scenarios*. If you are curious and would like to check the source code, you can visit its GitHub repository at `https://github.com/AdamTibi/OpenWeatherClient`.

You can install the NuGet package via the VS **graphical user interface** (**GUI**) by going to **Manage NuGet Packages…** and searching for `AdamTibi.OpenWeather` or via the .NET CLI, by going to the project directory and writing this:

```
dotnet add package AdamTibi.OpenWeather
```

The configuration is done, so now, we can modify the code.

Mapping feeling to temperature

Here is a simple method that will map the temperature in °C to a single word describing it:

```csharp
private static readonly string[] Summaries = new[]
{
    "Freezing", "Bracing", "Chilly", "Cool", "Mild",
    "Warm", "Balmy", "Hot", "Sweltering", "Scorching"
};
private string MapFeelToTemp(int temperatureC)
{
    if (temperatureC <= 0) return Summaries.First();
    int summariesIndex = (temperatureC / 5) + 1;
    if (summariesIndex >= Summaries.Length) return
        Summaries.Last();
    return Summaries[summariesIndex];
}
```

The output for 0 or less is `Freezing`, between 0 and 5 it is `Bracing`, then it is going to change every 5 degrees. 45 degrees onward, it is `Scorching`. Don't take my word for the output—we will unit test it. Imagine if we didn't!

Random weather API

I kept the random weather API, but I made it use the preceding MapFeelToTemp string, as follows:

```
[HttpGet("GetRandomWeatherForecast")]
public IEnumerable<WeatherForecast> GetRandom()
{
    WeatherForecast[] wfs = new
        WeatherForecast[FORECAST_DAYS];
    for(int i = 0;i < wfs.Length;i++)
    {
        var wf = wfs[i] = new WeatherForecast();
        wf.Date = DateTime.Now.AddDays(i + 1);
        wf.TemperatureC = Random.Shared.Next(-20, 55);
        wf.Summary = MapFeelToTemp(wf.TemperatureC);
    }
    return wfs;
}
```

This is a trivial API generating a random temperature and then making a summary out of the generated temperature. We are generating FORECAST_DAYS = 5 days, starting from the next day.

Running this project and hitting the Swagger UI output will give us this:

```
[
  {
    "date": "2021-11-26T22:23:38.6987801+00:00",
    "temperatureC": 30,
    "temperatureF": 85,
    "summary": "Hot"
  },
  {
    "date": "2021-11-27T22:23:38.7001358+00:00",
    "temperatureC": -15,
    "temperatureF": 6,
    "summary": "Freezing"
  },
...
```

You can see how random the output is, as the next day is hot but the day after is freezing.

Real weather API

The real weather API should make more sense. This is the newly added API:

```
[HttpGet("GetRealWeatherForecast")]
public async Task<IEnumerable<WeatherForecast>> GetReal()
{
    ...
    string apiKey = _config["OpenWeather:Key"];
    HttpClient httpClient = new HttpClient();
    Client openWeatherClient =
        new Client(apiKey, httpClient);
    OneCallResponse res = await
      openWeatherClient.OneCallAsync
        (GREENWICH_LAT, GREENWICH_LON, new [] {
            Excludes.Current, Excludes.Minutely,
            Excludes.Hourly, Excludes.Alerts },
            Units.Metric);
    ...
}
```

The method creates a HttpClient class for the sake of passing it to the *OpenWeather* Client class. It then fetches the API key and creates an *OpenWeather* Client class. To limit our scope, this will only forecast for Greenwich, London.

> **Important Note**
>
> The previous code is not clean and will be cleaned shortly in this chapter. If you really want to know the reason right now, it is instantiating (newing) the HttpClient and the Client classes in the controller, and this is not a good practice.

We are calling a RESTful API of *OpenWeather* called **OneCall**. This API returns today's weather and forecasts 6 consecutive days; this is good as we only need the next 5 consecutive days. The Swagger UI output of this one is shown here:

```
[
  {
    "date": "2021-11-26T11:00:00Z",
    "temperatureC": 8,
    "temperatureF": 46,
```

```
    "summary": "Chilly"
  },
  {
    "date": "2021-11-27T11:00:00Z",
    "temperatureC": 4,
    "temperatureF": 39,
    "summary": "Bracing"
  },
  ...
```

The best way to explain concepts is by example, so consider this test problem that will give you a firsthand experience of what DI is.

C to F conversion API

To have all the world come together and to keep everybody happy, we will add another method to convert °C to °F. We will have an API on our controller called ConvertCToF, and it looks like this:

```
[HttpGet("ConvertCToF")]
public double ConvertCToF(double c)
{
    double f = c * (9d / 5d) + 32;
    _logger.LogInformation("conversion requested");
    return f;
}
```

This API converts a temperature from °C to °F and logs every time this API is requested, for statistical purposes. You can invoke this API from Swagger UI as before, or invoke it from the browser like this:

https://localhost:7218/WeatherForecast/ConvertCToF?c=27

The output will look like this:

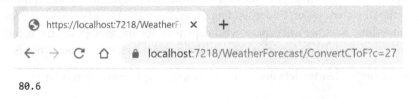

Figure 2.6 – Results of executing the ConvertCToF API from the browser

This is a **Unified Modeling Language** (**UML**) diagram showing what we have so far:

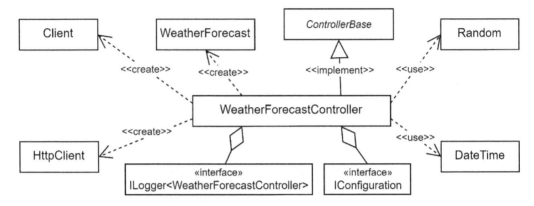

Figure 2.7 – UML diagram showing the WFA application

You can see all the changes in `WeatherForecastController.cs`; it is always in the `Ch02` source code in GitHub under a directory called `02-UqsWeather`.

The application is ready for critique, and I can tell you from now that the code you've just seen is not unit-testable. We can perform other categories of testing, but not unit testing, although it will be unit-testable by the end of this chapter. I invite you to open the project in VS and follow along, as we will implement exciting and important concepts.

Now that the project is ready, we need to set some basics right, and the first in the list is understanding dependency.

Understanding dependency

If your code does something useful, chances are your code depends on other code or another component, which in turn depends on another component. A clear understanding of the **dependency** terminology should give you a better grasp of unit testing and will definitely aid in having clearer conversations with your colleagues.

The plan in this section is to familiarize you with the concept of dependency, which should make understanding the DI pattern easier. Understanding dependency and DI are prerequisites for writing any serious unit testing. Next, we will explore what dependencies are, though when it comes to unit testing, we don't care for all dependencies, so we will define what a relevant dependency is.

Before we dive into dependencies, let's first define abstraction types and concrete types.

Abstractions and concrete types

To have you and me on the same channel, I will define the terminology to be used.

A concrete class is a class that can be instantiated; it could be something like this:

```
FileStream fileStream = new FileStream(…)
```

FileStream is a concrete type that can be instantiated and used directly in the code.

An abstraction type can be an abstract class or an interface. Examples of abstract classes are Stream, ControllerBase, and HttpContext. Examples of interfaces are IEnumerable, IDisposable, and ILogger.

I will be overusing these terms in the book, so it's worth defining them.

What is a dependency?

First, what it isn't: it is not equivalent to the same term used in UML.

In this book's context and when speaking with other developers within unit testing boundaries, it can be defined in this way: if class A *uses* a type of B where B is an abstraction type or a concrete class, then A has a dependency on B.

The term *uses* can be narrowed down to the following:

- B is passed to the constructor of A. Example from WFA: the logger is passed to the controller's constructor, which makes ILogger<WeatherForecastController> a dependency, as illustrated here:

```
public WeatherForecastController(
    ILogger<WeatherForecastController> logger, …)
```

- B is passed to a method in A, as follows:

```
public void DoSomething(B b) { …
```

- B has a static method that is called from a method in A. Example from WFA: DateTime. Now is called from GetRandom, which makes DateTime a dependency, as illustrated here:

```
wf.Date = DateTime.Now.AddDays(i + 1);
```

- B is instantiated anywhere inside A, whether instantiated in a method, in a field, or in a property. In the following example, HttpClient is instantiated in the code:

```
HttpClient httpClient = new HttpClient();
```

Based on this definition, we have all of the following as dependencies on `WeatherForecast Controller`:

- `Random`

- `DateTime`

- `Client`

- `HttpClient`

- `ILogger<WeatherForecastController>`

- `IConfiguration`

- `WeatherForecast`

Data transfer objects (**DTOs**) are not considered dependencies, although they look like concrete classes, but they act as a vehicle to carry data from one place to another. We will show an example of a DTO in the *The WeatherForecast class dependency* section.

Note that `record`, `record struct`, and `struct` usually follow the same concept as a DTO.

We will have more analysis of dependencies across *Part 1*, *Getting Started and Basics*, and *Part 2*, *Building an Application with TDD*. For an experienced TDD practitioner, spotting dependencies is second nature.

Dependency relevance

Dependencies lead our class to interact with components external to our code. A dependency is relevant for DI in the context of unit testing if it has a method or a property that might cause a side effect when triggered or when it leads to other behavior that is not very relevant to the class being tested.

This is an overloaded definition, and it is not meant to be all clear at this point. Examples will be provided from here until the end of *Part 2* to show when a dependency is relevant.

We care for pinpointing a dependency if we want to change its behavior when testing it. If `_logger. LogInformation` is writing to the disk, we want sometimes to change this behavior, especially when testing. As always, clarifying with examples is best, so in this section, we will demonstrate multiple examples and explain why they are relevant.

The logging dependency

Consider this `_logger` field:

```
private readonly ILogger<WeatherForecastController>
    _logger;
```

During the application lifespan, the `_logger` field might be triggered to write logs. Depending on the configuration of the logger, it might write logs in memory, in the console while debugging, in a log file on the disk, in the database, or on a cloud service such as **Azure Application Insights** or **Amazon CloudWatch**. We use the `_logger` field when we log in the `ConvertCToF` method, as follows:

```
_logger.LogInformation("conversion requested");
```

It is relevant because we have a side effect that will extend to other components in the system, and when unit testing at a later stage, we want to eliminate this side effect.

The configuration dependency

There is another field in the class, the `_config` field, as illustrated here:

```
private readonly IConfiguration _config;
```

The `_config` field is needed to get the API key from the configuration. It is passed through the constructor of the controller class, similar to the `_logger` field.

During runtime `_config` can load configuration based on configuration; pun not intended. Your configuration can be in the cloud, in `appsettings`, or in a custom format. We can see this dependency in use here:

```
string apiKey = _config["OpenWeather:Key"];
```

It is relevant as we need to go through the configuration to read the API key. Accessing configuration is also causing a side effect.

The HTTP dependency

Digging through the code, you find that we have instantiated `HttpClient` and used it in the code:

```
HttpClient httpClient = new HttpClient();
```

It is obvious that we have a dependency on **HyperText Transfer Protocol** (**HTTP**). Every time the method containing this code, `GetReal` API, is invoked, it issues an HTTP call.

Unlike the logging and configuration dependencies, where the dependency is built against an abstraction (`IConfiguration` and `ILogging<>`), `httpClient` is instantiated in the code—this makes what is called a **hard** or a **concrete dependency**.

We do care about the distinction between instantiating a dependency in code or passing it from outside, through the constructor. It'll be clear why later on.

It is relevant as we don't want to depend on the network while we are testing.

The OpenWeather client dependency

The OpenWeather client is a dependency on a dependency. It is a dependency itself and it is relying on the HTTP dependency, represented by `httpClient`, as well. You can see this in the following snippet:

```
Client openWeatherClient = new Client(apiKey, httpClient);
```

Also, this is another example of a concrete dependency as it is being instantiated inline.

It is relevant as we don't want to depend on HTTP (or the network) while we are testing.

The time dependency

Consider this line in the code:

```
wf.Date = DateTime.Now.AddDays(i + 1);
```

What is important here is the Now property. Now has code that will call the **operating system (OS)**, a dependency, asking for the current time. The Now property is static, as we can see here:

```
public static DateTime Now { get; }
```

The fact that this is static will make it slightly more difficult to deal with regarding DI, as we will see soon.

It is relevant as we want a predictable time during testing. Taking the current time will not lead to consistent results, as time is changing.

The randomness dependency

This is an example of depending on an algorithm to generate randomness:

```
wf.TemperatureC = Random.Shared.Next(-20, 55);
```

The Next method is a static method as well and it is calling the time in the background to generate a seed; also, it is depending on a *randomization algorithm*. We want to control the outcome so that we can test it.

It is relevant as we want predictable output.

The WeatherForecast class dependency

We are instantiating this class as a DTO, as we want to transfer the data from our method to the client. This data structure will be serialized into **JavaScript Object Notation (JSON)**. The code is illustrated here:

```
WeatherForecast[] wfs = new WeatherForecast[FORECAST_DAYS];
```

It is not relevant as this object does not cause a side effect and it just carries data.

If the code depends on abstractions and the objects are not instantiated in the class (the controller in the previous example), then this is generally good. If the code depends on concrete classes that are instantiated in the class, then we are not following best practices as we are violating one good **object-oriented programming (OOP)** practice: *depend on abstraction, not concrete*. This will be our next topic.

Depend on abstraction, not concrete

The title is popular advice in OOP best practices. This advice applies to two cases: the method signatures and the code inside the methods. We will explore both cases in this section.

Abstracted parameters in the method signature

When designing a method, including a class constructor, the advice is to check whether you can accept an abstracted type rather than a concrete type. As always, let's explain this with examples.

For an example of an abstract class, take the well-known `Stream` class from .NET, as illustrated in the following code snippet:

```
public abstract class Stream : …
```

A `Stream` object represents a sequence of bytes, but the class doesn't care about the physical source of the bytes—let it be from a file or from memory or others. This is the wisdom behind making it an abstract class.

We have `FileStream`, which inherits `Stream` as an example of a concrete class, as illustrated here:

```
public class FileStream : Stream
```

`FileStream` understands the specifications of reading a stream of bytes from a disk file.

We have also `MemoryStream`, which inherits `Stream` as another example of a concrete class, as illustrated here:

```
public class MemoryStream : Stream
```

Here is a UML diagram to summarize the relationship:

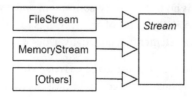

Figure 2.8 – Stream and its children

Having Stream as an abstract class opened the way for **decoupling** the implementation and simpler code. Consider this method from the System.Text.Json.JsonSerializer class, which accepts a parameter of the type Stream:

```
public static void Serialize (Stream utf8Json, object?
    value, ...);
```

This method converts the provided value to **Unicode Transformation Format 8 (UTF-8)**-encoded JSON text and writes it to the Stream class.

Because this method doesn't care for the way the Stream class is dealing with the underlying physical persistence, it is accepting the Stream parent abstract class as a parameter. If there is no abstraction, then you will have **multiple** overloads of the same method. Each one of them accepts a different stream type, like this (these overloads don't exist; they are hypothetical):

```
public static void Serialize (FileStream utf8Json, ...);
public static void Serialize (MemoryStream utf8Json, ...);
public static void Serialize (SqlFileStream utf8Json, ...);
public static void Serialize (BufferedStream utf8Json, ...);
More Stream overloads...
```

This is an example of the benefits of accepting an abstraction type as a method argument. Here is another example. Consider this code:

```
public static int Sum(int[] elements)
{
    int sum = 0;
    foreach (int e in elements) sum += e;
    return sum;
}
```

This method takes an array and returns the sum of its elements. At first glance, the method's signature looks fine, but if you think about it, this method forces the caller to convert any collection to an array before calling the method, which is an unnecessary conversion and a waste of performance as this method doesn't depend on the specific features of an array. It is just doing a foreach construct, which means it is accessing the array elements sequentially. Does it really need to accept an array?

Converting the signature parameter to IEnumerable<int>, which is an abstraction, would allow this method to accept a considerable number of concrete classes, as illustrated here:

```
public static int Sum(IEnumerable<int> elements)
```

You initially were only able to call this method with an `int []` array; now, it can be passed to any object of a class that implements `IEnumerable<int>`, and there are plenty of them. Here are a few:

```
int[] array = new int[] { 1, 2 };
List<int> list = new List<int> { 1, 2 };
Queue<int> queue = new Queue<int>();
queue.Enqueue(1);
queue.Enqueue(2);
// More collections
Sum(array);
Sum(list); // new benefit
Sum(queue); // new benefit
```

Going back to the WFA application, our controller constructor is already doing something right as it depends on abstractions, as illustrated in the following code snippet:

```
public WeatherForecastController(
    ILogger<WeatherForecastController> logger,
        IConfiguration config)
```

Always think of the common denominator abstraction type that satisfies the requirement to have your method as open as possible.

Direct instantiation of a dependency

We have just discussed using abstractions, when possible, in our method signature. This reduces coupling and increases the usability of the method. This section will extend the advice to the code.

If we are instantiating classes directly in the code, we depend on concrete objects. If we depend on concrete objects, then we cannot change their behavior at runtime. Let's take an example from our WFA application where we are instantiating the `Client` class in our method, as per this line of code:

```
Client openWeatherClient = new Client(apiKey, httpClient);
```

Then, whenever we use the `openWeatherClient` object, such as calling the `OneCallAsync` method, we will be firing an HTTP request over the network to a RESTful API on the *OpenWeather* end. This is good for the production code, but not good for testing the code; when we are testing, we want to eliminate this behavior.

> **Isolation**
>
> In this case, we can avoid the HTTP call and work around this using an *isolation framework* during testing. However, this is only kept as a last resort. We will explain what isolation frameworks are in *Chapter 3, Getting Started with Unit Testing*.

When testing the code, we don't want it to fire an HTTP request for many reasons, including the following:

- We have a limited number of calls that we can make per time—a quota.
- Our testing environment is behind a firewall that bans outbound traffic.
- The REST service on the other end of the network is temporarily down, so we will get a false-negative result that our test has failed.
- Calling a service over the internet is slow, compared to dealing with CPU and memory.

Can you see where we're going? The code works, but it is not testable in isolation of the HTTP calls.

> **Important Note**
>
> Some test categories should fire HTTP and reach the other end, such as integration tests. In the previous context, I meant tests that validate the business logic and do not test for connectivity—one of them is a unit test.

Instantiating concrete classes would not work if we were to unit test a piece of functionality. What we want to do during unit testing is to check if a *false attempt* to fire a call is made but not actually executed, and that would be enough. So far, the takeaway is that creating concrete classes in code is not compatible with DI and, accordingly, not compatible with unit testing.

The main solution to avoid instantiating classes in the business logic is DI, which is what we will see shortly.

Best practices recap

Our discussion in the *Depend on abstraction, not concrete* section boils down to these two examples of *do* and *do not*. Let's start with the bad or the do-not-do example, as follows:

```
public class BadClass
{
    public BadClass() {}
    public void DoSometing()
    {
        MyConcreteType t = new MyConcreteType();
```

```
        t.UseADependency();
    }
}
```

Here's the equivalent good class example:

```
public class GoodClass
{
    private readonly IMyClass _myClass;
    public GoodClass(IMyClass myClass)
        { _myClass = myClass; }
    public void DoSometing()
    {
      _myClass.UseADependency();
    }
    public void DoSometingElse(SecondClass second)
    {
      second.UseAnotherDependency();
    }
}
```

Here are the good practices:

- Having abstractions as parameters encourages decoupling and opens the method to accept more types.

- Depending on abstractions allows changing an object's behavior without changing the code in the class.

One question you would ask is this: *If I did not instantiate the objects that were passed to the constructor or the method at runtime, then who did? Surely somewhere along the line, some process has instantiated my dependencies and passed them to my class.* The answer to this question can be found in the next section.

Introducing DI

When I first learned how to do DI in code, I had a euphoria as if I had discovered a secret in software engineering; it is like *code magic*. We have been exploring dependencies in the previous sections and now, we are about to discover injecting these dependencies into our classes. The next step is explaining what DI is and using practical samples from the WFA application to make sure you are experimenting with a variety of scenarios. The best way to introduce DI is with a familiar example.

First example of DI

DI is all over any modern .NET code. In fact, we have one example right here in the ASP.NET template code:

```
public WeatherForecastController(
    ILogger<WeatherForecastController> logger)
{
    _logger = logger;
```

The `logger` object, which is a dependency, is injected into the controller when a new instance of the controller is created. There is nowhere in the controller that we are instantiating the `logger` class. It has been injected into the controller's constructor.

What does injection in this context mean? It means the ASP.NET framework found an incoming request that needs this controller to be instantiated. The framework realized that to create a new instance of `WeatherForecastController`, it needs to create an instance of a concrete class that implements `ILogger<WeatherForecastController>`, to do something similar to this:

```
ILogger<WeatherForecastController> logger = new
    Logger<WeatherForecastController>(…);
var controller = new WeatherForecastController(logger);
```

The constructor of the controller required an instance of a concrete class that implements `ILogger<WeatherForecastController>`, and the framework resolved that `Logger<>` implements `ILogger<>` and can be used as a parameter for the construction of the controller.

How did it resolve this? We will learn about this in the DI containers; what is important now is that it knew what to do in order to instantiate the controller class.

Now is the time to give every subject in our play a DI-related name, as follows:

- **DI container**: The software library that is managing the injection
- **Service**: The requested dependency (`ILogger<>` descendant object)
- **Client**: The class requesting the service (the controller, in the previous example)
- **Activation**: The process of instantiating the client
- **Resolution**: The DI container finding the right service required to activate the client

Testing an API

Let's dig deeper into DI with an example. Consider this test problem that will give you firsthand experience of what DI is. Take the `ConvertCToF` method we created earlier in our WFA application.

We want to do some tests for this method in order to validate whether the temperature conversion is done accurately. We have been given a few examples of °C and the equivalent °F for our tests, as follows:

1. -1.0 C = 30.20 F

2. 1.2 C = 34.16 F

To satisfy the tests, we want to use an old-school console application that will throw an exception if the conversion doesn't match the examples.

You can add the console application via the VS GUI or you can execute the following lines from the solution directory:

```
dotnet new console -o Uqs.Weather.TestRunner
dotnet sln add Uqs.Weather.TestRunner
dotnet add Uqs.Weather.TestRunner reference Uqs.Weather
```

This adds a new console application called `Uqs.Weather.TestRunner` to the existing solution, and references the existing ASP.NET Web API application. In VS, add this code to the `Program. cs` file of the console application:

```
using Microsoft.Extensions.Logging;
using Uqs.Weather.Controllers;
var logger = new Logger<WeatherForecastController>(null);
//fails
var controller = new WeatherForecastController(logger,
    null!);
double f1 = controller.ConvertCToF(-1.0);
if (f1 != 30.20d) throw new Exception("Invalid");
double f2 = controller.ConvertCToF(1.2);
if (f2 != 34.16d) throw new Exception("Invalid");
Console.WriteLine("Test Passed");
```

The code in the current format doesn't run as it fails at the `var  logger` line. We'll fix that in a moment, but let's analyze the code first. The code instantiates a controller, in the way we instantiate any class in .NET; then, it calls the `ConvertCToF` method and tries different values. If all values pass, then it will print **Test Passed**; otherwise, it will throw an exception.

To instantiate a `Logger<>` object, we need to pass to its constructor an object of `ILoggerFactory` type. If you pass `null`, it will fail at runtime. Besides, the bad news is that an instance of a concrete implementation of `ILoggerFactory` is not meant to be instantiated manually unless you are integrating a logging framework or handling a special case, and testing is not a special case! In brief, we cannot easily do this.

What if we try to pass to the controller's constructor two null values, and ignore creating a `Logger<>` object, like this:

```
var controller = new WeatherForecastController(null, null);
```

The problem is that if you pass a `null` value, your `_logger` object in the controller will be null and your code will fail at this line with the infamous `NullReferenceException` exception, as illustrated here:

```
_logger.LogInformation("conversion requested");
```

What we really want is just to instantiate the controller. We are not testing the logger; we want to pass to the constructor anything that will create an object from our controller, but the logger is standing in our way. It turns out that Microsoft has a class called `NullLogger<>` that does just that—getting out of the way! The documentation from Microsoft states "*Minimalistic logger that does nothing*".

With the enlightenment of this class, the first few lines of the code will look like this:

```
var logger = NullLogger<WeatherForecastController>
    .Instance;
var controller = new WeatherForecastController(logger, …);
```

We are getting a reference to `NullLogger<>` through the `Instance` field. When we call `_logger.LogInformation`, nothing will happen, which fits what we're looking for. If we run this console application now, we will get a **Test Passed** message.

> **Important Note**
> Testing methods via a console application is not the best practice for testing. Also, throwing exceptions and writing messages are not ideal for reporting failed and passed tests. The right way will be covered in the next chapter.

The constructor of the controller accepts an `ILogger<>` object, which gave us the flexibility of passing a `NullLogger<>` object as the latter implements `ILogger<>`, as illustrated here:

```
public class NullLogger<T> : Microsoft.Extensions.Logging
    .ILogger<T>
```

And the UML diagram of the logging classes looks like this:

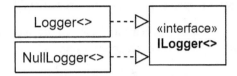

Figure 2.9 – UML of Logger<>, NullLogger<>, and ILogger<>

At this point, it is worth analyzing what we've done so far. Here's what we've achieved:

1. At runtime (when the APIs are launched), Logger<> was injected into the controller and it should be writing logs as expected.

2. At test time, we are not interested in the activities of logging; we are testing another scenario, so we've passed NullLogger<>.

3. We were allowed to inject into ILogger<> different types as ILogger<> is an interface, which is an abstraction. We would have failed to do this if our constructor expects a Logger<> type (the concrete type with no I).

In the first scenario, it was the **DI container** that injected the object at runtime. In the second scenario, this was us manually injecting a different logger for testing purposes. The annotated code in the following screenshot shows a summary of this section:

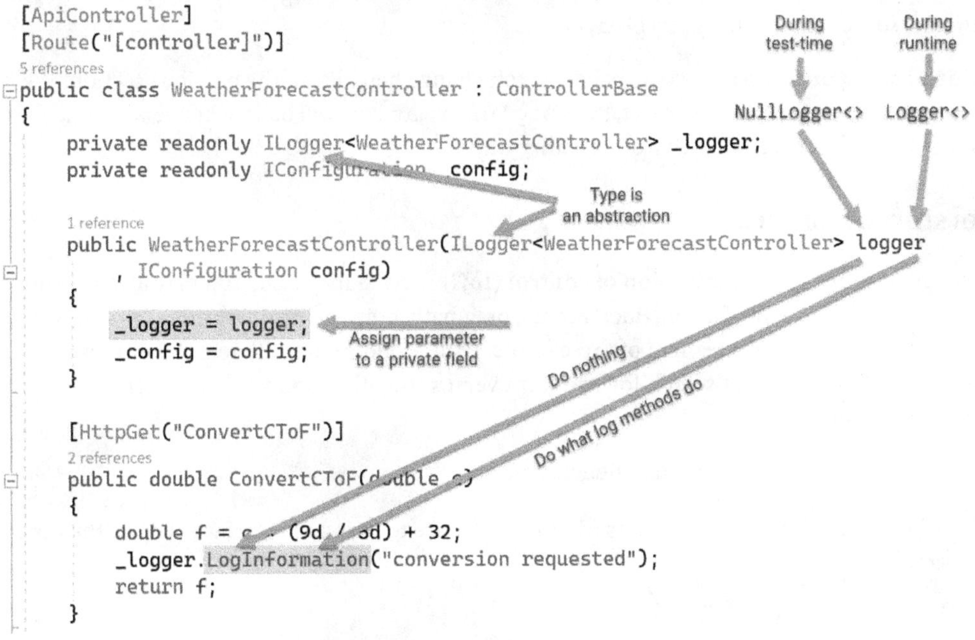

```
[ApiController]                                                During        During
[Route("[controller]")]                                       test-time     runtime
5 references
public class WeatherForecastController : ControllerBase
{                                                             NullLogger<>  Logger<>
    private readonly ILogger<WeatherForecastController> _logger;
    private readonly IConfiguration _config;
                                             Type is
    1 reference                             an abstraction
    public WeatherForecastController(ILogger<WeatherForecastController> logger
        , IConfiguration config)
    {
        _logger = logger;        Assign parameter
        _config = config;        to a private field
    }                                                    Do nothing

    [HttpGet("ConvertCToF")]                      Do what log methods do
    2 references
    public double ConvertCToF(double c)
    {
        double f = c * (9d / 5d) + 32;
        _logger.LogInformation("conversion requested");
        return f;
    }
}
```

Figure 2.10 – Annotated code showing DI at test time and at runtime

The conclusion here is that if our parameters use abstract types such as interfaces, `ILogger<>`-type interfaces, or abstract classes, we can open our classes for more reusability where DI can be utilized.

The `LogInformation` method is changing behavior based on the injected object, so it is acting as a seam. This drives us naturally to our next section about seams.

What are seams?

As an English term, a *seam* is where two pieces of fabric are stitched together. The term in a DI context resembles areas in the code where we can *change* the behavior without changing the code explicitly. We can point to the example from our previous convert method, shown here:

```
public double ConvertCToF(double c)
{
    double f = c * (9d / 5d) + 32;
    _logger.LogInformation("conversion requested");
    return f;
}
```

Take the `LogInformation` method. We want this method to write into some production instrument, but when we're testing, we want it to do nothing (if our test scenario is not about logging). We want to test other functionality, but `_logger.LogInformation` is standing in our way, trying to write somewhere, so we want to change its behavior.

`LogInformation` is a seam, as the behavior can change here. From the previous section, if we inject into the class a `Logger<>` object, then `LogInformation` will behave in one way, and if we inject `NullLogger<>`, it will behave in another way.

Inversion of control

You will often hear the term **inversion of control** (**IoC**) used to mean DI. You may also hear IoC container, as well, to mean a DI container. From a pragmatic point of view, you don't need to worry about the differences in the meaning of these terms. Practitioners have different definitions of IoC and how it relates to DI. Just search for one term versus the other and you'll find forums full of contradicting definitions.

Here are the common points that practitioners agree on:

- IoC is reversing the flow of events from the software to the **user interface** (**UI**) or the other way around.
- DI is a form of IoC.

DI is the most popular term and the most modern one. The term *IoC* is from a different era, is more generic, and has a less practical use, so I recommend using the term *DI*.

After all these examples, best practices, and definitions, I kept the best to last, which is the practical section of this chapter. This is how you can take all the previous literature and write useful code with it.

Using DI containers

A **DI container** is a library that injects a service into the client. A DI container provides extra functionality other than injecting dependencies, such as the following:

- Registering the classes that need to be injected (registering the services)
- Implementing how the services need to be instantiated
- Instantiating what has already been registered
- Managing the created service lifetime

Let's clarify a DI container role with an example from the previous code. We have the `logger` service being injected, but who is responsible for this?

There is a DI container called `Microsoft.Extensions.DependencyInjection` that will inject `_logger`. This happened in the first line of `Program.cs`, as illustrated here:

```
var builder = WebApplication.CreateBuilder(args);
```

This previous method call registers a default logger. Unfortunately, while we can see the code in the .NET source code, it is not obvious in our `Program.cs` source code. In fact, the previous line registers plenty of other services.

By adding a single line for experimentation, directly following the previous line in `Program.cs`, we can see how many registered services are created:

```
int servicesCount = builder.Services.Count;
```

This will give us 82 services. A few of these services are for logging-related activities. So, if you want to see what they are, you can have this line directly after the previous line:

```
var logServices = builder.Services.Where(_ =>
    x.ServiceType.Name.Contains("Log")).ToArray();
```

You can see here that we are filtering on any service that has the word Log as part of its name. If you have a breakpoint after this line and go to VS **Immediate Window** and type logServices, you can see a glimpse of all the registered log-related services, as illustrated in the following screenshot:

```
Immediate Window
logServices
{Microsoft.Extensions.DependencyInjection.ServiceDescriptor[10]}
    [0]: Lifetime = Singleton, ServiceType = {Name = "ILoggerFactory" FullName = "Microsoft.
    [1]: Lifetime = Singleton, ServiceType = {Name = "ILogger`1" FullName = "Microsoft.Exten
    [2]: Lifetime = Singleton, ServiceType = {Name = "ILoggerProviderConfigurationFactory" F
    [3]: Lifetime = Singleton, ServiceType = {Name = "ILoggerProviderConfiguration`1" FullNa
    [4]: Lifetime = Singleton, ServiceType = {Name = "LoggingConfiguration" FullName = "Micr
    [5]: Lifetime = Singleton, ServiceType = {Name = "ILoggerProvider" FullName = "Microsoft
    [6]: Lifetime = Singleton, ServiceType = {Name = "ILoggerProvider" FullName = "Microsoft
    [7]: Lifetime = Singleton, ServiceType = {Name = "LoggingEventSource" FullName = "Micros
    [8]: Lifetime = Singleton, ServiceType = {Name = "ILoggerProvider" FullName = "Microsoft
    [9]: Lifetime = Singleton, ServiceType = {Name = "ILoggerProvider" FullName = "Microsoft

Call Stack   Breakpoints   Command Window   Immediate Window   Output
```

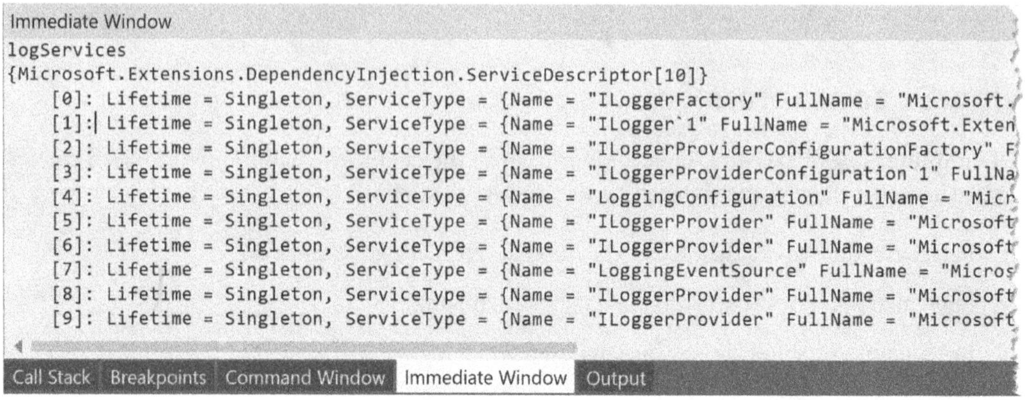

Figure 2.11 – Immediate window showing the registered logging-related services

The screenshot shows that we have 10 registered logging-related services. The one being injected for us at runtime is the second one (index number 1).

> **Note**
>
> You might get a different list of pre-registered services than this, depending on your ASP.NET version.

We will change our implementation in the controller to move everything to be dependency-injected and experiment with various scenarios of writing DI-ready code.

Container role

The container activities are being executed in the background by the DI container. A container is involved in booting up classes in your application, as illustrated in the following screenshot:

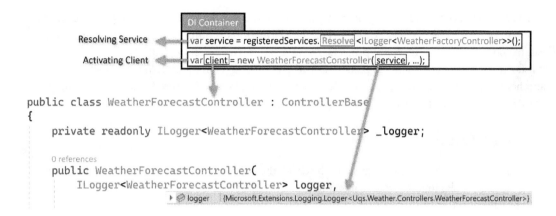

Figure 2.12 – Container in action (pseudo code)

The code of the **DI Container** box is a pseudo code. It is trying to summarize how the DI *resolves* the *service* required by the *client* from a list of already *registered* services. Then, the DI *activates* the client and passes it to the service. This is all happening at runtime.

The registration is an activity we will explore in many examples later on. In this scenario, there was an instruction that stated whenever an `ILogger<>` object is requested by the client, substitute it with a concrete class of the `Logger<>` type.

It is important to note that while the client is requesting an interface, the DI has been instructed earlier on how to construct a concrete class for this abstraction; the DI container knew earlier that to construct an `Ilogger<>` object, it needs to initialize a `Logger<>` object.

Third-party containers

We have been working so far with a *built-in* DI container that is automatically wired with new ASP.NET projects, and that is the `Microsoft.Extensions.DependencyInjection` Microsoft DI container, but this is not the only DI container available for .NET 6—there are other third-party options.

Microsoft has developed a DI container in recent years. Third-party containers gradually diminished in popularity in favor of the one shipped with .NET. Also, some frameworks did not make the leap with the introduction of .NET 5. Who's left strong today, with .NET 6, are **Autofac** and **StructureMap**. There are other containers supporting .NET 6, but they are not as popular.

If you are experienced in unit testing and you want more features that are not supported in `Microsoft.Extensions.DependencyInjection`, then have a look at other frameworks such as Autofac. But for non-monolithic, mid-size projects, I would recommend sticking with the Microsoft one as it is fairly supported and there are plenty of third-party plugin components. You can always swap to another framework at a later stage. My advice is not to spend valuable time choosing a DI container. Start with the Microsoft one until your requirements exceed it.

Service lifetime

When a service is registered to be passed to the client, the DI container has to decide about the lifetime of the service. The lifetime is the time interval from when the service is created until when it is released for garbage collection or disposed of.

The Microsoft DI container has three major lifetimes that you can specify when registering a service: **transient**, **singleton**, and **scoped lifetime scopes**.

Note that if the service implements the `IDisposable` interface, the `Dispose` method is invoked when the service is released. When a service is released, if it has dependencies, they are also released and disposed of. Next, we will explore the three major lifetimes.

Transient lifetime

Transient services are created every time they are injected or requested. The container simply creates a new instance for every request.

This is good in terms of not having to worry about thread safety or service state modification (by another requesting object). But creating an object for every request has adverse performance implications, especially when the service is in high demand, and activating it may not be cheap.

You will see an example of a transient service in the *Refactoring for DI* section later on.

Singleton lifetime

Singleton services are created once on the first client request and released when the application terminates. The same activated service will be passed to all requesters.

This is the most efficient lifetime as the object is created once, but this is the most dangerous one as a singleton service should allow concurrent access, which means it needs to be thread-safe.

You will see an example of a singleton service in the *Refactoring for DI* section later on.

Scoped lifetime

Scoped services are created once per HTTP request. They stay alive from the beginning of the HTTP request until the end of the HTTP response and they will be shared between clients.

This is good if you want one service to be used by several clients and the service applies to a single request only.

This lifetime is the least popular compared to the transient and the singleton lifetimes. Performance-wise, it sits in the middle between the transient and the singleton lifetimes. There is only one thread executing each client request at a given time, and because each request gets a separate DI scope, you don't have to worry about thread safety.

One popular example of using scoped services is using **Entity Framework's (EF's)** DB context object as scoped, which allows the request to share the same data and to cache data when required between clients.

Here is another example. Suppose you have a logging service that will allow the client to log, but it will only flush from memory to the destination media (say, saving to the database) when the HTTP request is over. Ignoring other conditions, this could be a candidate for a scoped lifetime.

We will have an example of a scoped lifetime in *Chapter 9, Building an Appointment Booking App with Entity Framework and Relational DB*.

Choosing a lifetime

If your concern is performance, then think of a singleton. Then, the next step is checking whether the service is thread-safe, either by reading its documentation or doing other types of investigation.

Then, fall down to scoped if relevant, and then fall down to transient. The safest option is always transient—if in doubt, then choose transient!

> **Important Note**
> Any class that gets injected into the singleton will become a singleton, regardless of the lifetime of the injected object.

Container workflow

Before we see some examples of service registration and lifetime, it's a good time to generalize our understanding of DI containers and look at a workflow diagram of the DI activation process:

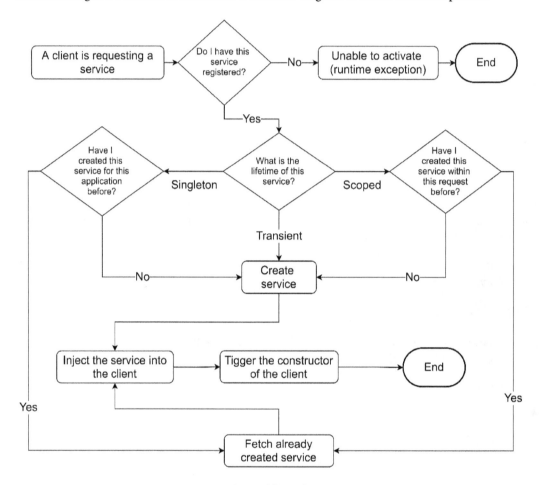

Figure 2.13 – The workflow of a DI container

In this diagram, it is clear that the DI container has two major concerns when activating a class, which are registration and lifetime.

Refactoring for DI

If you have done your DI right, then you are halfway with respect to implementing your unit tests. When writing a unit test, you will be thinking about how everything should be DI-ready.

There are certain factors that will determine how your services should be injected, as outlined here:

1. Does my seam belong to an abstracted method? In another way, does the method in question exist in an abstraction? This is the case with the `ILogger.LogInformation` method that we saw earlier, but we will cover this scenario again in more detail in the *Injecting the OpenWeather client* section.

2. Is my seam a static method? This will be covered in the *Injecting DateTime* and *Injecting the random generator* sections.

Injecting the OpenWeather client

One offending line is the `Client` class instantiation in `WeatherForecastController.cs`, as illustrated here:

```
string apiKey = _config["OpenWeather:Key"];
HttpClient httpClient = new HttpClient();
Client openWeatherClient = new Client(apiKey, httpClient);
OneCallResponse res =
    await openWeatherClient.OneCallAsync(…)
```

The sole purpose of accessing `_config` was to get the API key for `Client` and the sole purpose of instantiating `HttpClient` was to pass it to the constructor of `Client`. So, if we were to inject openWeatherClient, then the first two lines would not be needed.

Which method or property are we using from the to-be-injected class? The answer, by looking through the code, is `OneCallAsync` only. Then, what is the highest type (a class, an abstract class, or an interface) in the hierarchy of `Client` that has this member? To do this, hold the *Ctrl* button and click on the class name in VS, and you will find that `Client` implements `IClient`, as illustrated here:

```
public class Client : IClient
```

Then, hold *Ctrl* and click `IClient`, and you will find the following interface:

```
public interface IClient
{
    Task<OneCallResponse> OneCallAsync(decimal latitude,
        decimal longitude, IEnumerable<Excludes> excludes,
        Units unit);
}
```

Clearly, my implementation can depend on `IClient` rather than `Client`.

In the controller constructor, add `IClient` and add `_client` as a field, as follows:

```
private readonly IClient _client;
public WeatherForecastController(IClient client, …
{
    _client = client;
    …
```

The last step is doing the following modifications to these two lines:

```
Client openWeatherClient = new Client(apiKey, httpClient);
OneCallResponse res =
    await openWeatherClient.OneCallAsync(…);
```

Remove the first line as we are no longer instantiating `Client`, and modify your second line to use `_client` instead of the previous `openWeatherClient`. This will result in this code:

```
OneCallResponse res = await _client.OneCallAsync(…);
```

We have done all our modifications to the controller. What is left is registering with the DI container how to inject an object that matches `IClient` for our controller constructor. Let's run the project in its current state and we will get the following error:

```
System.InvalidOperationException: Unable to resolve service
  for type IClient' while attempting to activate
    'WeatherForecastController'
```

The DI container tried to look for a concrete class that implements `IClient` so that it can create it and pass it to the constructor of `WeatherForecastController`. We know that there is a concrete class called `Client` that implements `IClient`, but we have not told the DI container about it yet.

In order for the DI container to register a service, it requires two bits of information, as follows:

1. How to create the required service?
2. What is the lifetime of the created service?

The answer to *point 1* is we need to create an instance of `Client` whenever `IClient` is requested.

Point 2 is the trickier one. `Client` is a third-party class that is documented online. The first action is looking through the documentation to see whether it has a recommended lifetime and in this case, the documentation of `Client` specifies `Singleton` as the recommended one. In other cases where this is not documented, we have to figure it out in other ways. We will have more examples later on.

To register our dependency, in the `Program.cs` file, look for the comment provided by the `Add services to the container` template and add your code underneath it, as follows:

```
// Add services to the container.
builder.Services.AddSingleton<IClient>(_ => {
    string apiKey =
        builder.Configuration["OpenWeather:Key"];
    HttpClient httpClient = new HttpClient();
    return new Client(apiKey, httpClient);
});
```

Here, we are constructing `Client` in the same way we did it before. Once `Client` is first requested, only one instance will be created per application, and the same instance will be provided for all clients upon request.

Now, as we have finished with the DI of all dependencies required by the `GetReal` method, let's tackle the `Now` dependency in the `GetRandom` method.

Injecting DateTime

We are using `DateTime` in our `GetRandom` method and it is tricky to inject. Let's look at `DateTime` class usage in the code. We are using the following:

- `AddDays` method
- `Now` property, which returns a `DateTime` object

All this is clear in one line of code, shown here:

```
wf.Date = DateTime.Now.AddDays(i + 1);
```

The `AddDays` method is a method that relies on an arithmetic calculation of days, which can be verified by looking at the `DateTime` source code on GitHub, at `https://github.com/microsoft/referencesource/blob/master/mscorlib/system/datetime.cs`.

We don't have to worry about injecting it as it is not reaching an external dependency; it is just executing some C# code, or we might want to inject it to control how the `AddDays` method is being calculated. In our case here, injecting `AddDays` is not required.

The second point is the `Now` property. If we were to write a unit test that involves testing the value of `Now`, then we'd want to freeze it to a constant value to be able to test. At this stage, the picture of freezing it may not be clear, but it will be clearer when we unit test `GetRandom` in the next chapter.

We need to provide an injected Now property, but Now is a **static property**, as we can see here:

```
public static DateTime Now
```

Static properties (and methods) do not adhere to the same polymorphism principles that instance properties adhere to. So, we need to figure out another way to inject Now than what we used before.

The next code is preparing Now in a way suitable to work polymorphically. Create an interface like this one to act as an abstraction:

```
public interface INowWrapper
{
    DateTime Now { get; }
}
```

We will have our code depending on this abstraction type. Also, we will have to provide an implementation for a concrete NowWrapper class, so our code simply looks like this:

```
public class NowWrapper : INowWrapper
{
    public DateTime Now => DateTime.Now;
}
```

I have added two files under a directory called Wrappers in the project. I have added INowWrapper. cs and NowWrapper.cs under it.

Wrapper and Provider

Some developers like to have a Wrapper suffix for this category of types, and others like to use a Provider suffix such as NowProvider. I don't like to use the name Provider as it is already a design pattern and it might be misleading. My advice is to pick one convention and stay consistent.

As usual, we have two points to consider when registering a non-concrete type for injection, as follows:

1. How to create the required service?
2. What is the lifetime of the created service?

The first point is easy—we just instantiate the NowWrapper class. The second point depends on the DateTime.Now original property. Since I know that this is a web environment where multiple requests may be hitting my static property simultaneously, the first thing I would be checking is the popular .NET thread-safety topic. In other words, if this property is accessed simultaneously by multiple threads, would that lead to undetermined behavior?

Static members of DateTime, including the Now property, are written with thread safety in mind, so calling Now simultaneously should not lead to an undetermined behavior.

Given this is the case, then I can have my DI as a singleton. Let's register INowWrapper for injection. As with the previous example, add INowWrapper to the controller constructor, like this:

```
public WeatherForecastController(, INowWrapper nowWrapper, )
{
    _nowWrapper = nowWrapper;
...
```

Replace DateTime.Now with _nowWrapper.Now, as follows:

```
wf.Date = _nowWrapper.Now.AddDays(i + 1);
```

And lastly, register your dependency in the Program.cs file, using the following code:

```
builder.Services.AddSingleton<INowWrapper>(_ =>
    new NowWrapper());
```

This means that when the first INowWrapper instance is requested, the DI container will instantiate it and keep it for the lifetime of the application.

Injecting the random generator

The random number generator is unpredictable by design; otherwise, it wouldn't be random! There is a problem in unit testing it if it is not DI-injected, because the unit tests should be testing against a fixed (determined) value. Let's look at the offending line here:

```
wf.TemperatureC = Random.Shared.Next(-20, 55);
```

Shared is a static method, so we have the same issue that we had in the previous task with Now. First, we need to determine thread safety. There is no definite mention in the Next documentation if it is thread-safe; on the contrary, claims online mention that it is not thread-safe. So, the safest option here is to assume that it is not thread-safe. Here, we can wrap the entire class or the particular method. I will choose to wrap the entire class in case we need it later to use another method from the Random class. Let's write our interface, as follows:

```
public interface IRandomWrapper
{
    int Next(int minValue, int maxValue);
}
```

And here, we have the concrete class implementing it:

```
public class RandomWrapper : IRandomWrapper
{
    private readonly Random _random = Random.Shared;
    public int Next(int minValue, int maxValue)
        => _random.Next(minValue, maxValue);
}
```

Add this as usual to the controller constructor and replace the code in GetRandom with this:

```
wf.TemperatureC = _randomWrapper.Next(-20, 55);
```

I did change the behavior slightly in the class; initially, it was creating a new Random instance every time we call Next, but now it is creating one _randomWrapper per requesting class.

As our Next class implementation depends on the thread-unsafe _random.Next, then our class is not thread-safe as well. So, when injecting it, we cannot inject it as a singleton; we have to inject it as a transient, so our Program.cs code looks like this:

```
builder.Services.AddTransient<IRandomWrapper>(_ =>
    new RandomWrapper());
```

This might have worked as a AddScoped registration method, but the documentation is insufficient for me to decide, and transient is always the safest.

You can now run the application, and from the Swagger UI, execute both APIs to make sure that everything is working as expected.

The DI changes that we have done are all in the Ch02 source code in GitHub under a directory called 03-UqsWeather.

Lifelike DI scenario

The most common scenario for using DI is with unit testing, though I have seen it being used elsewhere to change the behavior of a certain component at runtime. Take the case where you want to change a functionality of a system based on a configuration, and another case where you want to change a system behavior per hosting environment. Consider the next example of **load-testing** our WFA application.

Using DI as a load-testing example

A common **non-functional requirement** (**NFR**) for critical systems is load testing. Load testing is an artificial simulation of calls to a system to measure how it handles high volumes of concurrent calls. For our WFA, load testing would look like this:

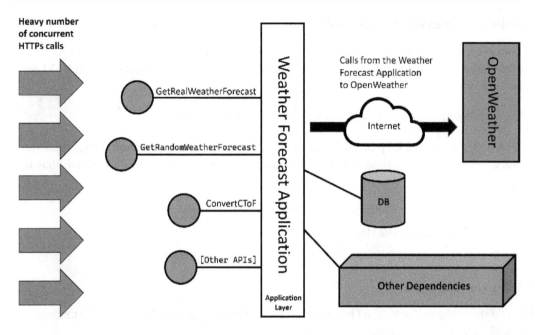

Figure 2.14 – WFA under load testing

A load-testing framework would start the test by issuing a pre-agreed number of calls to the APIs and measuring response times and failures. In turn, the APIs will exert loads on their dependencies.

The complete WFA might have multiple dependencies, but what we are interested in specifically, in this example, is the *OpenWeather* API that we are calling in the background. If we are to load-test the WFA application, we will, by design, issue a heavy number of calls to *OpenWeather* and this *should not* be the case, for many reasons. Here are a few:

- Consuming the number of allocated calls quota
- Contractual agreement against load-testing their system through yours
- Being banned for an exorbitant number of calls in a short period
- Ethical reasons, as this might affect their overall service quality

Unless your system requires specifically to load-test with the third party connected and you have an agreement with the third party to do so, I wouldn't do it.

What can we do to work around this and conduct our load test without calling *OpenWeather*?

A solution could be to add a configuration key to the WFA. When this key is true, we want every call in all our application to *OpenWeather* to return a stubbed response (canned response). More about dummy, mock, stub, and fake will be discussed in the next chapter. For now, we will refer to this type of response as a stubbed response.

Enabling an OpenWeather stubbed response

Let's enable a stub response representing OpenWeather. Where do we start? I would directly look for the *seam* that is causing the call to OpenWeather. It is in our `WeatherForecastController` class, as illustrated here:

```
OneCallResponse res = await _client.OneCallAsync (…)
```

What we need to do is to keep the previous code the same but make this method change behavior by not going over the network and instead return some saved value when under a load test. Here is the plan to achieve this:

1. Add a configuration to denote load testing.
2. Add a stubbed response class.
3. Register a condition to swap responses based on the configuration.

Adding configuration

We want the configuration to be off by default unless we explicitly set it *on*. In your `appsettings.json` file, add the following code:

```
"LoadTest": {
   "IsActive" : false
}, …
```

And in our `appsettings.Development.json` file, add the same configuration, but set it to `true`. This should result in `true` when you load the application locally.

Adding the stub class

`OneCallAsync` is a method on the `IClient` interface. If you look at the code, we are passing the `client` object, which becomes `_client`, as an argument to the constructor. Here is where we can do some magic—we need to pass to the constructor our stubbed implementation of `IClient`, then figure out a way to pass it through the constructor.

Add a class called `ClientStub` to the root of your project to hold the implementation of our stubbed `IClient` interface, as follows:

```
public class ClientStub : IClient
{
    public Task<OneCallResponse> OneCallAsync (
        decimal latitude, decimal longitude,
        IEnumerable<Excludes> excludes, Units unit)
```

```
    {
        const int DAYS = 7;
        OneCallResponse res = new OneCallResponse();
        res.Daily = new Daily[DAYS];
        DateTime now = DateTime.Now;
        for (int i = 0; i < DAYS; i++)
        {
            res.Daily[i] = new Daily();
            res.Daily[i].Dt = now.AddDays(i);
            res.Daily[i].Temp = new Temp();
            res.Daily[i].Temp.Day =
                Random.Shared.Next(-20, 55);
        }
        return Task.FromResult(res);
    }
}
```

IClient is defined in the NuGet package for the *OpenWeather* client. It has one method to implement OneCallAsync. I looked for the used properties and generated a 7-day fake forecast. Note that you might need to make a full stub in other scenarios.

Now, both Client and ClientStub implement IClient, as per this diagram:

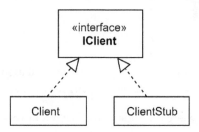

Figure 2.15 – IClient, Client, and ClientStub relationship

Now comes the step that developers forget to do often: registering the service. Remember that every time you forget to register a service, you are not alone.

Updating IClient registration

We are going to use our DI container to decide when to inject an instance of `Client` and when to inject an instance of `ClientStub`. In `Program.cs`, modify the initial register of `IClient` so that it looks like this:

```
builder.Services.AddSingleton<IClient>(_ => {
    bool isLoad =
    bool.Parse(builder.Configuration["LoadTest:IsActive"]);
    if (isLoad) return new ClientStub();
    else
    {
        string apiKey =
            builder.Configuration["OpenWeather:Key"];
        HttpClient httpClient = new HttpClient();
        return new Client(apiKey, httpClient);
    }
});
```

Whenever an instance of `IClient` is requested, the DI container will decide what to inject, `ClientStub` or `Client`, based on the configuration.

We now have the stubbing implementation completed and ready to run. Have a look at the output of the `GetReal` method when you run the project. You will notice you are getting the stubbed version if you've enabled load testing.

Points to note

We have seen, dare I say it, a beautiful way of swapping implementation. While this example is small and contained, the implementation will shine more in larger projects. Consider these points:

- **Separation of concerns** where the code to load different versions is taken away from the controller class to the registration section.

- The developer will not need to worry about or remember to do additional implementation when passing `IClient` to new controllers.

Similar to this scenario, you can use DI whenever a swap of implementation is required under certain conditions.

This scenario is located in the `Ch02` source code in GitHub under a directory called `04-UqsWeather`.

Method injection

You have seen across the chapter that we have been injecting parameters through the constructor. There is another less popular form of injection, called **method injection**. This is an example from the WFA controller:

```
public double ConvertCToF(double c,
    [FromServices] ILogger<WeatherForecastController>
        logger)
{

    double f = c * (9d / 5d) + 32;
    logger.LogInformation("conversion requested");
    return f;
}
```

Notice the FromServices attribute. This instructs the DI container to inject a dependency into a method in the same way it is injected into a constructor. Obviously, this is not needed in a constructor.

You would use method injection when you have multiple methods in a class. One of them uses a special service. The benefit here is a cleaner class constructor and a bit of performance saving because the class—for example, the controller—might be instantiated, but the injection service would have a chance of not being used. So, there is a performance waste in injecting it but not using it.

In this example case, the logger was only used in the ConvertCToF method, so it can be moved from the constructor to the method. It needs to be injected only when ConvertCToF, not the controller, is instantiated to serve any other method.

Best practices recommend classes with a single responsibility. This leads to related methods with related services, so you won't find method injection as a popular pattern, but method injection is there if you need it.

Property injection

Property injection is injecting a service into a property on a class. This is not supported by Microsoft containers, but it is supported by third-party containers.

I have seen this used with legacy systems where a DI container is gradually being introduced and code changes are at a minimum. However, I have never seen or used this in a greenfield application.

I trust that it was not added to the Microsoft container, as it is not popular and not encouraged.

Service locator

Every container comes with or integrates with a **service locator**. A service locator finds and activates a registered service. So, the DI container registers a service and the service locator resolves what is already registered. Here is a typical pattern of using a service locator:

```
public class SampleClass
{
    private readonly IServiceProvider _serviceProvider;
    public SampleClass(IServiceProvider serviceProvider)
    {
        _serviceProvider = serviceProvider;
    }
    public void Method()
    {
        MyClass myClass =
            _serviceProvider.GetService<IMyClass>();
        myClass.DoSomething();
    }
}
```

`IServiceProvider` is an abstraction that supports service location. It can be injected into a class like any other service. Notice when we called the `GetService` method that it got us whatever is registered with `IMyClass`.

Obviously, you could have done the same thing by injecting `IMyClass` into the constructor, and it is even better to do so. You can see this being done here:

```
public SampleClass(IMyClass myClass)
```

But there are situations where you will want to avoid injection and prefer to use a service locator. This is more often used in legacy applications where DI is not fully implemented.

Using a service locator in code will complicate your unit tests, so it is better avoided, and some practitioners would consider using it as an anti-pattern.

Summary

This is a long chapter, I admit, but my defense is that it has plenty of examples to cover many real-life DI scenarios. Also, DI automatically encourages good software engineering practices, so we had to include the relevant practices. If you were to develop TDD-style, you would spend around 10% of your coding time doing DI-related tasks, and I hope this chapter did the right job and added to your knowledge.

DI is mainly used with unit tests, so without it, DI may feel less interesting. The next chapter, *Getting Started with Unit Testing*, will use the WFA application that we refactored here, and hopefully, you will appreciate further this design pattern.

Further reading

To learn more about the topics discussed in the chapter, you can refer to the following links:

- *IoC*: `https://martinfowler.com/bliki/InversionOfControl.html`
- *DI in ASP.NET Core*: `https://docs.microsoft.com/en-us/aspnet/core/fundamentals/dependency-injection`

3

Getting Started with Unit Testing

Unit testing is the core of TDD and a prerequisite for practicing it. I want to briefly go through the minimal necessary theory and focus more instead on familiarizing you with the tools and techniques that unit testing practitioners utilize in their daily code.

Here, you will learn how to write unit tests that cover moderate coding scenarios. In *Part 2, Building an Application with TDD*, of this book, we will take the knowledge acquired in this chapter to a higher level and use it in a lifelike fashion.

In the previous chapter, we built the **weather forecasting application (WFA)** and made it a **dependency injection (DI)**-ready. We will use this application in this chapter as the basis for learning about unit testing. If you are not familiar with DI and DI containers, I recommend starting with *Chapter 2, Understanding Dependency Injection by Example*, first.

In this chapter, we will do the following:

- Introduce unit testing
- Explain the structure of a unit testing project
- Analyze the anatomy of a unit test class
- Discuss the basics of xUnit
- Show how SOLID principles and unit testing are related

By the end of the chapter, you will be able to write basic unit tests.

Technical requirements

The code for this chapter can be found at the following GitHub repository:

`https://github.com/PacktPublishing/Pragmatic-Test-Driven-Development-in-C-Sharp-and-.NET/tree/main/ch03`

Introducing unit testing

As a TDD practitioner, you will be writing much more unit test code than production code (the regular application code). Unlike other test categories, unit tests will dictate some architectural decisions of your application and enforce DI.

We won't dwell on long definitions. Instead, we will demonstrate unit testing with a plethora of examples. In this section, we will discuss the xUnit unit testing framework and the unit test structure.

What is unit testing?

Unit testing is testing a behavior while swapping real dependencies with test doubles. Let me back up this definition with an example from `WeatherForecastController` in the WFA:

```
private readonly ILogger<WeatherForecastController>
    _logger;
public double ConvertCToF(double c)
{
    double f = c * (9d / 5d) + 32;
    _logger.LogInformation("conversion requested");
    return f;
}
```

This method converts Celsius to Fahrenheit and logs every call. Logging is not the concern here because this method's concern is the conversion.

The *behavior* of this method is *converting input degrees from Celsius to Fahrenheit*, and the logging *dependency* here is accessed through the `_logger` object. At runtime, we are injecting a `Logger<>` that will be writing to a physical medium, but we possibly want to eliminate the writing side effect when testing.

Based on the earlier definition, we need to *swap the real dependency* that `_logger` uses at runtime with its test double counterpart and test the conversion behavior. We will show how to do this later on in this chapter.

Take another example from the same class:

```
private readonly IClient _client;
public async Task<IEnumerable<WeatherForecast>> GetReal()
{
    ...
    OneCallResponse res = await _client.OneCallAsync(...
    ...
}
```

The behavior of this method is getting the real weather forecast and passing it to the caller. The _client object here represents the OpenWeather dependency. This method's behavior is *not* about interacting with the details of the RESTful protocol of the OpenWeather API or the HTTP protocol. This is handled by _client. We need to swap the real dependency, Client, that _client uses at runtime, and replace it with one that is suitable for testing (we call this a **test double**). I will show how this is done, in many ways, in *Chapter 4, Real Unit Testing with Test Doubles*.

At this stage, the concept is still cryptic, I know; just bear with me, and we will start expanding gently. In the next section, we will discuss unit testing frameworks. We will need this to unit test the preceding examples and the WFA.

Unit testing frameworks

.NET 6 has three major unit testing frameworks. The most popular one is **xUnit**, which we will use across this book. The other two are **NUnit** and **MSTest**:

- *NUnit* is an open source library. It started as a port from Java's JUnit framework and was then completely rewritten. You will still encounter it in legacy projects, but the majority of today's projects start with xUnit.

- *MSTest* is Microsoft's unit testing framework that gained popularity because it used to be shipped with Visual Studio and no extra effort was needed to install it, especially since NuGet did not exist back then. It became open source in version 2, and it was always lagging behind NUnit and then xUnit in terms of features.

- *xUnit* is an open source project that was started by developers from NUnit. It is feature-rich and in constant development.

> **Note**
> The term **XUnit** is an umbrella term for different languages' unit test frameworks, such as **JUnit (Java)**, **NUnit (.NET)**, **xUnit (.NET)**, and **CUnit (C language)**. This should not be confused with the library name *xUnit*, which is a .NET unit test library, where the founders picked an already-taken and confusing name.

Learning one framework and then switching to another should take no time, as they are similar, and you just need to figure out the terminology used by the specific framework. Next, we will add an xUnit project to the solution to unit test the WFA.

Understanding test projects

xUnit templates come as part of VS. We will show how to add an xUnit project using the **.NET CLI** approach. At this stage, if you have not opened the companion source code that is ported from *Chapter 2, Understanding Dependency Injection by Example*, to this chapter, I encourage you to do so.

Adding xUnit via the CLI

Currently, we have a solution with one ASP.NET Core project. Now, we want to add the unit tests library to our solution. To do so, create a new xUnit project called Uqs.Weather.Tests.Unit in a directory with the same name, and use .NET 6.0:

```
dotnet new xunit -o Uqs.Weather.Tests.Unit -f net6.0
```

Add the newly created project to the solution file:

```
dotnet sln add Uqs.Weather.Tests.Unit
```

Now, we have two projects in our solution. As the unit test project will be testing the ASP.NET Core project, the unit test project should have a reference to the ASP.NET Core project.

Add a project reference from Uqs.Weather.Tests.Unit on Uqs.Weather:

```
dotnet add Uqs.Weather.Tests.Unit reference Uqs.Weather
```

We now have our solution fully constructed via the CLI. You can see the full interaction here:

```
Command Prompt                    ×    +  ∨

C:\R\TDD\Book\ch03>dotnet new xunit -o Uqs.Weather.Tests.Unit -f net6.0
The template "xUnit Test Project" was created successfully.

Processing post-creation actions...
Running 'dotnet restore' on C:\R\TDD\Book\ch03\Uqs.Weather.Tests.Unit\Uqs.Weather.Tests.Unit.csproj...
  Determining projects to restore...
  Restored C:\R\TDD\Book\ch03\Uqs.Weather.Tests.Unit\Uqs.Weather.Tests.Unit.csproj (in 799 ms).
Restore succeeded.

C:\R\TDD\Book\ch03>dotnet sln add Uqs.Weather.Tests.Unit
Project `Uqs.Weather.Tests.Unit\Uqs.Weather.Tests.Unit.csproj` added to the solution.

C:\R\TDD\Book\ch03>dotnet add Uqs.Weather.Tests.Unit reference Uqs.Weather
Reference `..\Uqs.Weather\Uqs.Weather.csproj` added to the project.

C:\R\TDD\Book\ch03>
```

Figure 3.1 – Creating a new xUnit project in a solution via the CLI

We now have a project to contain our unit tests.

Test project naming convention

You have noticed that we have appended .Tests.Unit to the original project name, so the unit test project became Uqs.Weather.Tests.Unit. This is a common convention in naming test projects.

This convention extends to other testing projects, such as integration testing and Sintegration testing, to be discussed in the *More testing categories* section in *Chapter 4*. You might also have the following:

- Uqs.Weather.Tests.Integration

- Uqs.Weather.Tests.Sintegration

The wisdom behind this convention is that you can look at your list of projects and rapidly find the test projects related to one production code project ordered next to each other, as follows:

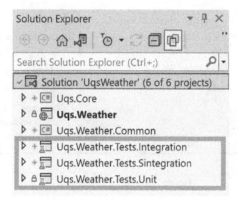

Figure 3.2 – Ordered unit test projects

The convention also helps in targeting all your test projects in the continuous integration, which will be covered in *Chapter 11, Implementing Continuous Integration with GitHub Actions*, in case you wanted to run all categories of tests. Here is an example: Uqs.Weather.Tests.*.

Running the sample unit test

The xUnit template comes with a sample unit test class called UnitTest1.cs that has a sample unit test method with the following content:

```
using Xunit;
namespace Uqs.Weather.Tests.Unit;
public class UnitTest1
```

```
{
    [Fact]
    public void Test1()
    {
    }
}
```

This has a single unit test called `Test1` that is empty and does nothing at the moment. To check that the xUnit framework and the integration with VS does work, you can try executing this single test.

From the VS menu, select **Test | Run All Tests** or similarly use the *Ctrl + R, A* keyboard shortcut. This will execute all your tests in the project (which is currently one test), and you will have the following tool, known as **Test Explorer**.

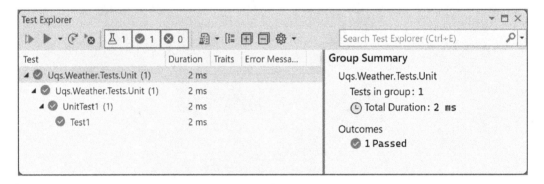

Figure 3.3 – Test Explorer

The hierarchy displayed here is **Project Name | Test Class Namespace | Test Class Name | Test Method Name**.

If you are the CLI kind of person, you can go to the solution directory with the command prompt and execute the following:

```
dotnet test
```

This is what you may get:

```
C:\R\TDD\Book\ch03>dotnet test
  Determining projects to restore...
  All projects are up-to-date for restore.
  Uqs.Weather -> C:\R\TDD\Book\ch03\Uqs.Weather\bin\Debug\net6.0\Uqs.Weather.dll
  Uqs.Weather.Tests.Unit -> C:\R\TDD\Book\ch03\Uqs.Weather.Tests.Unit\bin\Debug\net6.0\Uqs.Weather.Tests.Unit.dll
Test run for C:\R\TDD\Book\ch03\Uqs.Weather.Tests.Unit\bin\Debug\net6.0\Uqs.Weather.Tests.Unit.dll (.NETCoreApp,Version=v6.0)
Microsoft (R) Test Execution Command Line Tool Version 17.0.0
Copyright (c) Microsoft Corporation.  All rights reserved.

Starting test execution, please wait...
A total of 1 test files matched the specified pattern.

Passed! - Failed:      0, Passed:      1, Skipped:      0, Total:      1, Duration: < 1 ms - Uqs.Weather.Tests.Unit.dll (net6.0)

C:\R\TDD\Book\ch03>
```

Figure 3.4 – CLI dotnet test results

I have seen **Test Explorer** used more in day-to-day TDD-style development than the CLI. The CLI is useful for running the whole solution or for continuous integration and automated runs.

Test Explorer

Test Explorer comes with VS. Additionally, xUnit adds a few libraries that allow Test Explorer and VS to interact with xUnit tests. There are third-party providers that have more advanced test runners. One of them is *JetBrains ReSharper Unit Test Explorer*. We have everything ready to start writing unit test code.

Unit test class anatomy

When we unit test, we tend to write a **unit test class** that is targeting a parallel production class – one test class against one production class.

Applying this concept to our WFA project, our production class is WeatherForecastController and the unit test class is going to be called WeatherForecastControllerTests. So, rename the UnitTest1 sample class to WeatherForecastControllerTests.

> **Tip**
>
> You can set your text cursor anywhere within the class name in the source code (in the previous case, it was UnitTest1) and hit *Ctrl + R, R* (hold *Ctrl* then press *R* quickly twice). Type the new name WeatherForecastControllerTests and hit *Enter*. This will also rename the file if the **Rename symbol's file** checkbox is ticked.

Next, we will see how to organize our unit test class and its methods.

Class naming convention

I found the most commonly used convention is calling the unit test class name the same as the production code class name, appending the `Tests` suffix. For example, the `MyProductionCode` test class counterpart would be `MyProductionCodeTests`.

When practicing TDD, you will need to switch between the test class and its counterpart production code class multiple times in a short period. Naming them using this pattern allows you to find the test and its related counterpart easily, or vice versa. It also clarifies the relationship between the two classes.

Test methods

Each test class contains methods that test pieces of functionality, known as units, from the production code class. Let's take the example of testing the `ConvertCToF` method.

Test example 1

Part of the requirements that we have is testing the conversion with a single decimal point accuracy. So, let's consider one testing case by taking a zero degree (0.0 C) and testing if the method is returning 32.0 F. To do that, we can delete the `Test1` method in the unit tests class and replace it with the following:

```
[Fact]
public void ConvertCToF_0Celsius_32Fahrenheit()
{
    const double expected = 32d;
    var controller = new WeatherForecastController(
        null!, null!, null!, null!);
    double actual = controller.ConvertCToF(0);
    Assert.Equal(expected, actual);
}
```

This code initializes the production code class, calls the method under test, and then assesses the results of the test with our expectations.

`Fact` is an attribute that makes a method a unit test. `Assert` is a static class that has useful methods for comparing expected results to actual results. Both `Fact` and `Assert` are part of the xUnit framework.

Run this test with Test Explorer using *Ctrl + R, A*, and the test will yield the following screen:

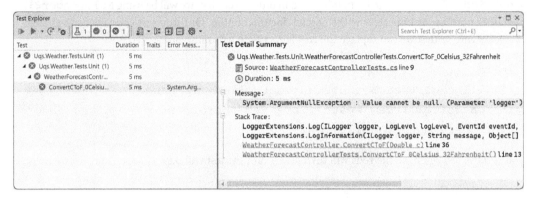

Figure 3.5 – Fail output in Test Explorer

If we look inside the controller, we find that VS has a red sign to map the route that led to this failure:

```
1 reference | ⊗ 0/1 passing   ⬅
public WeatherForecastController(ILogger<WeatherForecastController> logger,
    IClient client, INowWrapper nowWrapper, IRandomWrapper randomWrapper)
{
    _logger = logger;
    _client = client;
    _nowWrapper = nowWrapper;
    _randomWrapper = randomWrapper;
}


[HttpGet("ConvertCToF")]
1 reference | ⊗ 0/1 passing   ⬅
public double ConvertCToF(double c)
{
    double f = c * (9d / 5d) + 32;
    _logger.LogInformation("conversion requested");
    return f;
}
```

Figure 3.6 – VS showing the failed test route

It is apparent from the error message what is causing `ArgumentNullException`:

```
_logger.LogInformation("conversion requested");
```

This is expected, as we have passed the `logger` parameter from the unit test as `null`. We want _ `logger.LogInformation` to do nothing, and in order to do that, we will be using `NullLogger<>`, which does nothing, as indicated by the official documentation. Our unit test code needs to change to the following so that we can replace the real logger with a dummy one:

```
var logger =
    NullLogger<WeatherForecastController>.Instance;
var controller = new WeatherForecastController(
    logger, null!, null!, null!);
```

If you run the test again, all the reds will turn green, and the test will pass.

Test example 2

To test another input and output for the method, we can add more unit tests to the class and follow the same test method name pattern. We can have the following:

```
public void ConvertCToF_1Celsius_33p8Fahrenheit() {…}
…
public void ConvertCToF_Minus1Celsius_30p2Fahrenheit() {…}
```

But, there is a succinct solution for avoiding writing a similar unit test for every value combination, as follows:

```
[Theory]
[InlineData(-100 , -148)]
[InlineData(-10.1,  13.8)]
[InlineData(10    ,  50)]
public void ConvertCToF_Cel_CorrectFah(double c, double f)
{
    var logger =
        NullLogger<WeatherForecastController>.Instance;
    var controller = new WeatherForecastController(
        logger, null!, null!, null!);
    double actual = controller.ConvertCToF(c);
    Assert.Equal(f, actual, 1);
}
```

Notice that we are using `Theory` rather than `Fact`. Every `InlineData` will act as a single unit test. You can even eliminate *example 1* and make it an `InlineData` attribute. No need to say that `Theory` and `InlineData` are attributes from xUnit.

You can go ahead and run the tests.

Other examples are covered in *Chapter 1, Writing Your First TDD Implementation*, similar to the previous ones in this chapter, so you could have a look for more clarity.

Examples 1 and *2* target a simple method, `ConvertCToF`, that has a single dependency, `_logger`. We will cover more sophisticated testing scenarios after learning about test doubles in *Chapter 4, Real Unit Testing with Test Doubles*. In reality, your production code will be more complicated than a simple conversion method and will contain multiple dependencies, but there is a first step for everything.

Naming convention

Unit test method names follow a popular convention: `MethodUnderTest_Condition_Expectation`. We have seen this convention used earlier. Here are more hypothetical examples:

- `SaveData_CannotConnectToDB_InvalidOperationException`
- `OrderShoppingBasket_EmptyBasket_NoAction`

This book contains plenty of other examples that should clarify this convention further.

The Arrange-Act-Assert pattern

The previous test method, and generally all unit test methods, follows a similar pattern:

1. Create a state, declare some variables, and do some preparations.
2. Call the method under test.
3. Assert the actual results against expectations.

Practitioners decided to give these three stages the following names:

Arrange, **Act**, and **Assert** (AAA).

They mark the code with a comment to show the stages and emphasize the separation. According to this, we can write one of the previous test methods as follows:

```
[Fact]
public void ConvertCToF_0Celsius_32Fahrenheit()
{
    // Arrange
    const double expected = 32d;
    var controller = new WeatherForecastController(…);

    // Act
```

```
        double actual = controller.ConvertCToF(0);

        // Assert
        Assert.Equal(expected, actual);
    }
```

Notice the comments that were added to the code.

> **Important Note**
>
> Some teams dislike having the separation by writing comments. Instead, they choose a different way to mark AAA, for example, by leaving a single line space between each section.

The AAA practice is more than a convention. It makes the method easier to read on-the-fly. It also emphasizes that there should be only one *Act* in a unit test method. Consequently, a unit test, based on best practices, is not supposed to have more than a single AAA structure.

Using VS code snippets

Every unit test is going to have the same structure. VS allows you to cut down on writing the same structure with the help of **code snippets**. I have included a code snippet file for unit testing in the CodeSnippets directory in this chapter source code. It is called aaa.snippet. You can open it and view/edit its content via a regular text editor (not a word processor).

To use this snippet on Windows, copy aaa.snippet to this directory (choosing the right VS version):

```
%USERPROFILE%\Documents\Visual Studio 2022\Code Snippets\Visual
C#\My Code Snippets
```

Once this is copied, in your unit test class, type aaa, then hit the *Tab* button, and you will get the following generated code:

```
[Fact]
public void Method_Condition_Expectation()
{
    // Arrange

    // Act

    // Assert
}
```

Rather than speaking more about having a single AAA in your unit test, we will demonstrate this across this book to illuminate the style that seniors use in writing unit tests.

Now that we had an overview of the class anatomy and the unit test method structure, we will explore the unit test class's counterpart: the system under test.

System under test

A unit test is meant to test a single functionality of a production code. Each unit test class has a production code counterpart that is being tested. We refer to the production code being tested as the **system under test (SUT)**. You can see here an illustration of what the SUT is:

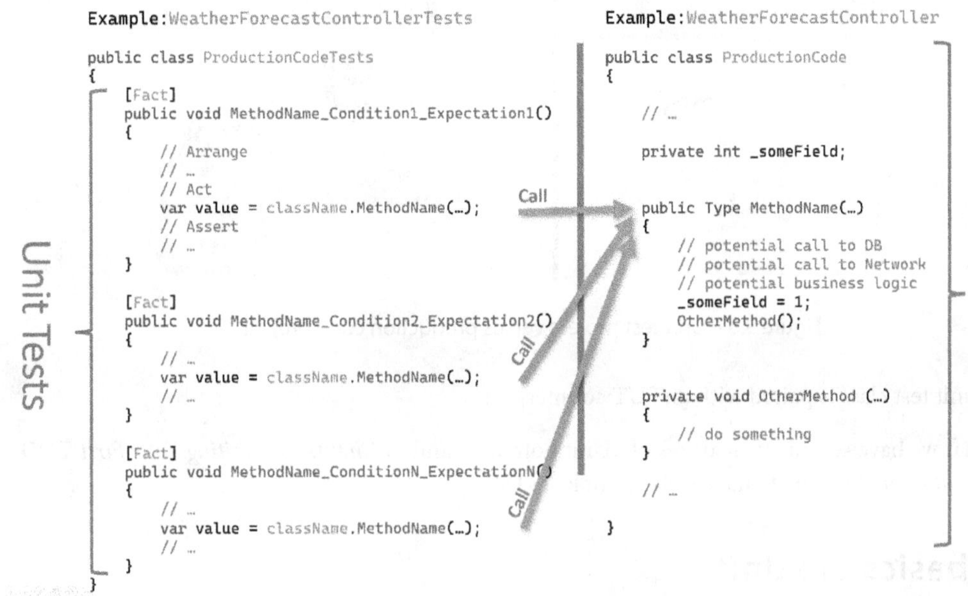

Figure 3.7 – Unit tests operating against the SUT

The term SUT is the most dominant one, but you might find others refer to it as **class under test (CUT)**, **code under test** (CUT – yes, it is the same acronym), or **method under test (MUT)**.

The term *SUT* is used in developers' conversations, and it is also commonly used in the code to make it clear what is being tested, like this:

```
var sut = new ProductionCode(…);
```

It is important to understand the SUT of your unit test class. You will gradually notice as your project grows that you will have a pattern being formed, as follows:

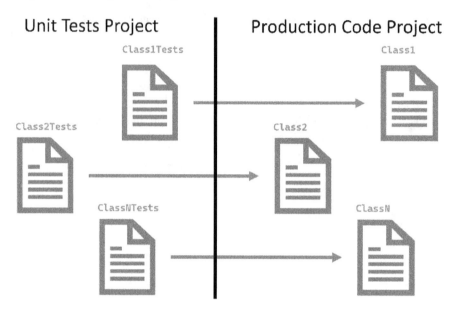

Figure 3.8 – Unit test project versus production code project

Every unit test class is paired with an SUT counterpart.

Now that we have seen a few features of xUnit both here and in *Chapter 1, Writing Your First TDD Implementation*, it is time to have a closer look at xUnit.

The basics of xUnit

xUnit provides the hosting environment for your tests. One important feature of xUnit is that it is AAA-convention friendly. It also integrates with the VS IDE and its Test Explorer.

Extensive examples using xUnit appear naturally in this book. However, it is worth dedicating a few sections to discussing the principal features of this framework.

Fact and theory attributes

In your test project, any method that is decorated with Fact or Theory will become a test method. Fact is meant for a non-parametrized unit test, and Theory is for a parametrized one. With Theory, you can add other attributes, such as InlineData, for parametrization.

> **Note**
>
> VS will give you a visual indication above the method name that you can run the methods decorated with these attributes, but sometimes it doesn't until you run all the tests.

Running the tests

Each unit test will *run independently* and instantiate the class. The unit tests *do not share* each other's states. So, a unit test class runs differently than a normal class. Let me elaborate with a sample code, shown here:

```
public class SampleTests
{
    private int _instanceField = 0;
    private static int _staticField = 0;

    [Fact]
    public void UnitTest1()
    {
        _instanceField++;
        _staticField++;
        Assert.Equal(1, _instanceField);
        Assert.Equal(1, _staticField);
    }

    [Fact]
    public void UnitTest2()
    {
        _instanceField++;
        _staticField++;
        Assert.Equal(1, _instanceField);
        Assert.Equal(2, _staticField);
    }
}
```

The previous unit tests do pass. Notice that while I am incrementing `_instanceField` in both test methods, the value of `_instanceField` is not shared between the two methods, and every time xUnit is instantiating a method, all my class is instantiated again. This is why the value is reset back to 0 before every method execution. This characteristic of xUnit promotes a unit test principle known as **no interdependency**, which will be discussed in *Chapter 6, The FIRSTHAND Guidelines of TDD*.

On the other hand, the static field was shared between the two methods and its value has changed.

> **Important Note**
>
> While I have used both instance and static fields to illustrate the distinctive behavior of a unit test class, I want to emphasize that using a static `read-write` field in a unit test is an anti-pattern because this breaks the *no interdependency* principle. In general, you should have no common `write` fields in unit test classes, and fields are better to be marked with the `readonly` keyword.

Instead, if the same methods are part of a regular code class (not a unit test class) and both are called, we expect to find the value of `_instanceField` incremented to 2, but this was not the case here.

Assert class

`Assert` is a static class, and it is part of xUnit. This is how the official documentation defines the `Assert` class:

Contains various static methods that are used to verify0 that conditions are met.

Let's have a quick overview of some of the methods of `Assert`:

- `Equal (expected, actual)`: These are a series of overloads that will compare the expectations to actuals. You have seen a few examples of `Equal` in *Chapter 1, Writing Your First TDD Implementation*, and in this chapter.

- `True (actual)`: Rather than using `Equal` to compare two objects, you can use this one where relevant to promote readability. Let's clarify this with an example:

  ```
  Assert.Equal(true, isPositive);
  // or
  Assert.True(isPositive);
  ```

- `False (actual)`: The opposite of the previous method.

- `Contains (expected, collection)`: A group of overloads that check for the existence of a single element in a collection.

- `DoesNotContain (expected, collection)`: The opposite of the previous method.

- `Empty(collection)`: This verifies that a collection is empty.
- `Assert.IsType<Type>(actual)`: This verifies whether an object is of a certain type.

As there are more methods, I encourage you to visit the official xUnit site to have a look, or to do what most developers do: write `Assert` in a unit test class, and type a *dot* after it to trigger IntelliSense and view the displayed methods.

The methods of `Assert` will communicate with the test runner, such as Test Explorer, to report back the result of the assertions.

Record class

The `Record` class is a static class that records exceptions so that you can test whether your method is throwing or not throwing the right exception. This is an example of one of its static methods, which is called `Exception()`:

```
public static System.Exception Exception(Action testCode)
```

The previous code returns the exception that is thrown by `Action`. Let's take this example:

```
[Fact]
public void Load_InvalidJson_FormatException()
{
    // Arrange
    string input = "{not a valid JSON";

    // Act
    var exception = Record.Exception(() =>
        JsonParser.Load(input));

    // Assert
    Assert.IsType<FormatException>(exception);
}
```

Here, we are checking whether the `Load` method will throw `FormatException` if it is presented with an invalid JSON input.

This was a summary of xUnit's functionality, and this should get you started writing basic unit tests.

Applying SOLID principles to unit testing

The SOLID principles are highly covered and advertised on the web and in books. Chances are that this is not the first time that you've heard or read about them. They are a popular interview question as well. SOLID principles stand for the following:

- Single-responsibility principle
- Open-closed principle
- Liskov Substitution principle
- Interface Segregation principle
- Dependency Inversion

In this section, we are interested mostly in the relationship between the SOLID principles and unit testing. While not all the principles have strong ties with unit testing, we will cover all of them for completion.

Single-responsibility principle

The **single-responsibility principle (SRP)** is about having each class with one responsibility only. This will lead it to have one reason to change. The benefits of this approach are as follows:

- **Easier to read and understand classes**:

  The classes will have fewer methods, which should cause less code. Its interface will have fewer methods as well.

- **Less rippling effect when changing a feature**:

  There are fewer classes to change, which would lead to an easier change.

- **Less probability of change, which means fewer potential bugs**:

  More code means more potential bugs, and changing code would also lead to potential bugs. Having less code in the first place means fewer code changes.

Example

The SRP is not an exact science, and the challenge is being able to decide what *responsibility* is. Every developer has their own view. The next example illustrates the idea.

Let's assume you have created your own file format called *ABCML* to solve a particular problem, as the existing file formats (such as JSON, XML, and others) do not satisfy your specific need. A set of classes with each having a single responsibility could be as follows:

- A class to validate whether the content of the file is in the right structure

- A class to export ABCML to a generic format

- A class that inherits the generic ABCML export to support exporting to JSON, and another class that supports exporting to XML

- A class that represents a node in your ABCML

- More classes

You can see how I split responsibilities into individual classes, although there is no single design for having a single responsibility.

SRP and unit testing

Naturally, when doing unit testing, you think of a single responsibility for a class, and you call your unit test class the same name with the `tests` suffix. So, if you are thinking of testing the validation of the ABCML file format, you might have `ABCMLValidationTests`.

In your unit test class, each unit test targets a single behavior in your SUT. Those behaviors combined lead to a single responsibility.

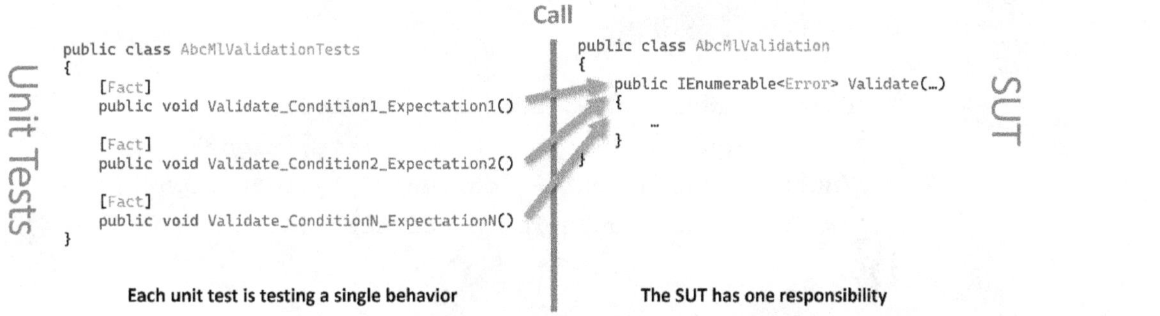

Figure 5.8 – Multiple single behavior tests targeting a single responsibility

The previous figure shows multiple tests, each test is focused on a single behavior, and they are targeting one responsibility: *validation*. On the right, there is one method, but this is only for illustration, as you might have multiple public methods and you can still have a single responsibility.

In *Chapter 6, The FIRSTHAND Guidelines of TDD*, we will introduce a guideline known as the *single-behavior guideline*. This guideline works with TDD and unit tests to encourage the SRP.

Open-closed principle

The **open-closed principle (OCP)** is about preparing your class to be inheritable (having it open) so that any feature addition could just inherit this class without modifying it (having it closed).

The essence of this principle is to minimize unnecessary changes every time a new feature is added.

Example

Let's take an example that will make this clearer. Assume we have created a library to do arithmetic calculations. Let's start by *not being OCP-compliant*, as shown here:

```
public interface IArithmeticOperation {}
public class Addition : IArithmeticOperation
{
    public double Add(double left, double right) =>
        left + right;
}
public class Subtraction : IArithmeticOperation { … }
public class Calculation
{
    public double Calculate(IArithmeticOperation op,
        double left, double right) =>
        op switch
        {
          Addition addition => addition.Add(left, right),
          Subtraction sub => sub.Subtract(left, right),
          //Multiplication mul => mul.Multiply(left,right),
          _ => throw new NotImplementedException()
        };
}
```

The `Calculate` method in the preceding code will have to change every time we add a new `ArithmeticOperation`. If we want to add the multiplication operation, per the commented line, as a feature at a later stage, then the `Calculate` method will need to change to accommodate the new feature.

We can make this implementation more OCP-compliant by eliminating the need to change the `Calculate` method every time a new operation is added. Let's see how this can be done:

```
public interface IArithmeticOperation
{
    public double Operate(double left, double right);
}
public class Addition : IArithmeticOperation
{
    public double Operate(double left, double right) =>
        left + right;
}
public class Subtraction : IArithmeticOperation { … }
// public class Multiplication : IArithmeticOperation { … }
public class Calculation
{
    public double Calculate(IArithmeticOperation op,
        double left, double right) =>
            op.Operate(left, right);
}
```

The previous example leveraged polymorphism to stop the `Calculation` method from being changed every time a new operation is added. You can see from the commented line how a new multiplication operation can be added. This is a more OCP-compliant approach.

> **Note**
>
> While I had all the classes and interfaces listed together here and on the GitHub code, I did this for illustration, as they are usually separated into their own files. So, with the OCP, you also reduce the chance of changing the file and make it easier on the source control level to understand what changed.

OCP and unit testing

Unit testing protects changes in any class by making sure that a change does not inadvertently break an existing feature. The OCP and unit testing work hand in hand. So, while the OCP reduces the chance of avoidable changes, the unit testing adds an additional protection layer when a change is made by verifying business rules.

Liskov substitution principle

The **Liskov substitution principle** (**LSP**) states that an instance of a child class must replace an instance of the parent class without affecting the results that we would get from an instance of the base class itself. A child class should be a true representation of its parent class.

Example

We shall use an academic type of example that will make the concept easier to understand. Let's take the following example:

```
public abstract class Bird
{
    public abstract void Fly();
    public abstract void Walk();
}
public class Robin : Bird
{
    public override void Fly() => Console.WriteLine("fly");
    public override void Walk() =>
        Console.WriteLine("walk");
}
public class Ostrich : Bird
{

    public override void Fly() =>
        throw new InvalidOperationException();
    public override void Walk() =>
        Console.WriteLine("walk");

}
```

In the previous code, and according to LSP, Ostrich should not have inherited Bird. Let's rectify the code to comply with the LSP:

```
public abstract class Bird
{
    public abstract void Walk();
}
public abstract class FlyingBird : Bird
{
    public abstract void Fly();
```

```
}
public class Robin : FlyingBird
{
    public override void Fly() => Console.WriteLine("fly");
    public override void Walk() =>
        Console.WriteLine("walk");
}
public class Ostrich : Bird
{
    public override void Walk() =>
        Console.WriteLine("walk");
}
```

We have changed the inheritance hierarchy by introducing a new intermediary class called FlyingBird for compliance with the LSP.

LSP and unit testing

Unit testing has no direct impact on the LSP, but the LSP is mentioned here for completion.

Interface segregation principle

The **interface segregation principle** (**ISP**) states that child classes should not be forced to depend upon interfaces that they do not use. Interfaces should be smaller so that whoever is implementing them can mix and match.

Example

I always find the way collections implement in .NET is the best example to explain this principle. Let's look at how List<T> is declared:

```
public class List<T> : ICollection<T>, IEnumerable<T>,
    IList<T>, IReadOnlyCollection<T>, IReadOnlyList<T>, IList
```

It is implementing six interfaces. Each interface contains a limited number of methods. List<T> provides a huge number of methods, but the way it does that is by selecting multiple interfaces, with each interface adding a few of the methods.

One method that List<T> exposes is GetEnumerator(). This method comes from the IEnumerable<T> interface; actually, it is the only method on IEnumerable<T>.

By having small interfaces (interfaces of few and related methods), as in this example, `List<T>` was able to choose what it needs to implement, no more and no less.

ISP and unit testing

Unit testing has no direct impact on the ISP, but the ISP is mentioned here for completion.

Dependency inversion principle

The **dependency inversion principle** (**DIP**) states that high-level modules should not depend on low-level modules. Both should depend on abstractions. Abstractions should not depend on details. Details should depend on abstractions. In other words, the DIP is a principle that promotes loose coupling between classes by using abstractions and DI.

Example

Chapter 2, Understanding Dependency Injection by Example, is focused on this topic, and it is rich with examples of changing the code to enable DI.

DIP and unit testing

There is a tight relationship between the DIP and unit testing. Real unit tests cannot function without DI. In fact, the effort spent on making everything injectable and having the proper interface designs for classes without interfaces promotes the DIP as a byproduct.

You can see that the SRP and the DIP are promoted by unit testing. So, while you are increasing your production quality, your design quality is improving as a result. There is no argument that unit testing requires effort, but part of this effort is already paid into your design quality and code readability.

Summary

In this chapter, we touched on basic unit testing-related topics, and we went through several examples.

If I was to categorize unit testing experience from 1 to 5, with level 1 being a beginner and 5 being an expert, this chapter should get you to level 2. Fear not! After going through the rest of the book, where more realistic examples will come, you will be at level 4, so I am glad you have made it so far. Keep going!

Is this book going to take me to level 5? I hear you asking. Well, unit testing is not a sprint, it is a marathon; it takes years of practice to get to that level, and only getting your hands dirty in unit testing will get you there.

We also covered the relationship between SOLID principles and unit testing to show you the big picture and how everything fits nicely together.

In this chapter, I have deliberately avoided examples that require a deep understanding of test doubles, so as to introduce you to unit testing in a gentle way. However, in reality, most unit tests will require test doubles. Let's move forward to a more realistic spectrum and dive into this concept in the next chapter.

Further reading

To learn more about the topics discussed in the chapter, you can refer to the following links:

- *Walkthrough: Create a code snippet*: `https://docs.microsoft.com/en-us/visualstudio/ide/walkthrough-creating-a-code-snippet`

- *xUnit*: `https://xunit.net`

4

Real Unit Testing with Test Doubles

A unit test differentiates itself from other test categories by using **test doubles**; actually, you would rarely see a unit test without a test double.

There is a lot of confusion on the web about what that means. My aim in this chapter is to clarify this term so that you can use it in the right context and give you as many explained examples of the topic so you feel confident in selecting the right test double for the test at hand.

In this chapter, we will:

- Explain the concepts and usage of test doubles
- Discuss more testing categories

By the end of the chapter, you will understand what is special about unit testing and will be able to use test doubles to start writing realistic unit tests.

Technical requirements

The code for this chapter can be found at the following GitHub repository:

```
https://github.com/PacktPublishing/Pragmatic-Test-Driven-Development-
in-C-Sharp-and-.NET/tree/main/ch04
```

Understanding and using test doubles

You would rarely write a unit test without using a **test double**. Think of the term *double* in the same sense as a Hollywood stunt, where a stunt takes the place of a real actor in some situations. *Test double* is an umbrella term for an object that is used to replace a dependency with a test equivalent (double) for the sake of testing a SUT. They are meant to satisfy one or more of the following requirements:

> **Requirement 1**: Enable the test code to compile.
>
> **Requirement 2**: Eliminate side effects according to the unit test requirements.
>
> **Requirement 3**: Embed a canned (predetermined) behavior that relates somehow to the real behavior.
>
> **Requirement 4**: Take a note of and verify the activities that were exerted on a dependency within a unit test (we will name this requirement later as *spying*).

We will be referring to these four conditions when we discuss individual test double types, so you may want to bookmark this section.

Do you want your method to call the payment gateway and execute a transaction while being unit tested? Do you want to call a third-party API that costs money while you are unit testing? Do you even want to go over the HTTP while you are testing? *Hint*: You don't want, and you shouldn't.

Let's understand the different types of test doubles that can satisfy the four conditions mentioned earlier.

Types of test doubles

There are five major types of test doubles—each one is meant to satisfy one or more of the four requirements mentioned earlier. When unit testing, you may use zero or more types of test doubles to satisfy your tests.

Next, we will discuss dummies, stubs, mocks, and fakes. These four types of test doubles are commonly used with TDD. The fifth type is isolation, which is not used with TDD and is only mentioned here for completeness.

Dummies

Dummies are a straightforward type of test double; in fact, I used them earlier. Dummies are meant to be passed to your SUT to let the code compile. If dummies are used in code, then they are supposed to do nothing. Consider the controller class of the WFA application and the `ConvertCToF` method:

```
// Constructor
public WeatherForecastController(
    ILogger<WeatherForecastController> logger,
    IClient client, INowWrapper nowWrapper,
```

```
    IRandomWrapper randomWrapper)
...
public double ConvertCToF(double c)
{
    double f = c * (9d / 5d) + 32;
    _logger.LogInformation("conversion requested");
    return f;
}
```

To test the ConvertCToF, we have to instantiate a WeatherForecastController class. The constructor expects multiple objects to be passed in to instantiate the controller class: logger, client, nowWrapper, and randomWrapper. But the ConvertCToF is only using _logger. Also, we did not want to test the side effect of _logger as we are testing another behavior. For this reason, we decided to use a NullLogger<>. We can pass all dummies to our controller, like this:

```
var logger =
    NullLogger<WeatherForecastController>.Instance;
var sut = new WeatherForecastController(logger, null, null,
    null);
```

When logger is used, it does nothing, and the other null values are just passed to make the code compile. In this case, logger and the null values are acting as dummy test doubles.

Creating *intelligent* test doubles when dummies can be used can complicate and blur the intention of your unit test, so use dummies when you can.

Dummies satisfy the first and the second test double requirements that we've mentioned previously. They allow the code to compile and also create objects that do nothing when called.

Stubs

Stubs are classes that respond with canned, pre-coded behavior. They are easy to write, easy to read, and won't require a special framework. The caveat is that they are harder to maintain than mocks. Take the GetReal() method of the WFA controller:

```
OneCallResponse res = await _client.OneCallAsync
    (GREENWICH_LAT, GREENWICH_LON, new[] {
        Excludes.Current, Excludes.Minutely,
        Excludes.Hourly, Excludes.Alerts }, Units.Metric);


WeatherForecast[] wfs = new WeatherForecast[FORECAST_DAYS];
```

```
for (int i = 0; i < wfs.Length; i++)
{
    var wf = wfs[i] = new WeatherForecast();
    wf.Date = res.Daily[i + 1].Dt;
    double forecastedTemp = res.Daily[i + 1].Temp.Day;
    wf.TemperatureC = (int)Math.Round(forecastedTemp);
    wf.Summary = MapFeelToTemp(wf.TemperatureC);
}
return wfs;
```

We are using _client, a dependency service, and calling OneCallAsync to retrieve the weather data from *OpenWeather*. This saves the results in the res object. OneCallResponse is not what we want to return to the GetReal() API consumer. Instead, we want to present the consumer with a simple output collection of the type WeatherForecast[]. For this reason, we have a mapping process that takes the data coming from _client.OneCallAsync and maps it to WeatherForecast[].

In the preceding code, the point that links the mapping process to *OpenWeather* is the OneCallAsync call:

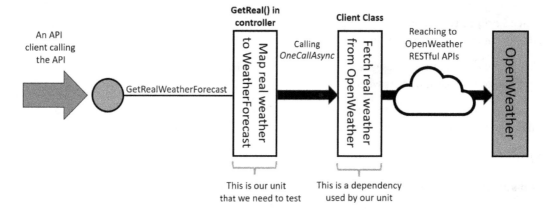

Figure 4.1 – The unit we need to test

We want to swap the implementation of OneCallAsync with our own stubbed implementation to avoid calling the real RESTful API because the unit that we are testing is the mapping business logic. Luckily, we can swap the implementation using *polymorphism*. This can be done through implementing IClient ourselves by creating a concrete class called ClientStub and writing our own OneCallAsync method. Our final design looks like this:

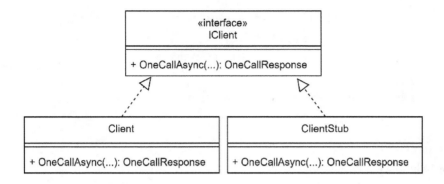

Figure 4.2 – Client and ClientStub implementing IClient

Let's build our stub class:

```
public class ClientStub : IClient
{
    private readonly DateTime _now;
    private readonly IEnumerable<double> _sevenDaysTemps;

    public ClientStub(DateTime now,
                      IEnumerable<double> sevenDaysTemps)
    {
        _now = now;
        _sevenDaysTemps = sevenDaysTemps;
    }

    public Task<OneCallResponse> OneCallAsync(
        decimal latitude, decimal longitude,
        IEnumerable<Excludes> excludes, Units unit)
    {
        const int DAYS = 7;
        OneCallResponse res = new OneCallResponse();
        res.Daily = new Daily[DAYS];
        for (int i = 0; i < DAYS; i++)
        {
            res.Daily[i] = new Daily();
            res.Daily[i].Dt = _now.AddDays(i);
```

```
            res.Daily[i].Temp = new Temp();
            res.Daily[i].Temp.Day =
                _sevenDaysTemps.ElementAt(i);
        }
        return Task.FromResult(res);
    }
}
```

Notice the following in the preceding code:

- `ClientStub` implements `IClient`, and it should provide an implementation for `OneCallAsync` to satisfy the contract.

- The constructor allows the user of the class to provide the `DateTime` and the seven days temperature.

- The `OneCallAsync` method has a made-up, stubbed implementation that generates a `OneCallResponse` return.

Now that we have written the class, we can put it into practice. There are several test criteria that we want to test. Here is the first test with the first criteria:

```
public async Task
    GetReal_NotInterestedInTodayWeather_WFStartsFromNextDay()
{
    // Arrange
    const double nextDayTemp = 3.3;
    const double day5Temp = 7.7;
    var today = new DateTime(2022, 1, 1);
    var realWeatherTemps = new double[]
        {2, nextDayTemp, 4, 5.5, 6, day5Temp, 8};
    var clientStub = new ClientStub(today,
        realWeatherTemps);
    var controller = new WeatherForecastController(
        null!, clientStub, null!, null!);

    // Act
    IEnumerable<WeatherForecast> wfs = await
        controller.GetReal();
```

```
      // Assert
      Assert.Equal(3, wfs.First().TemperatureC);
}
```

Notice that we are deciding what day it is. This is our way of freezing the day so that the test can execute anytime. We are also deciding what the weather is going to be for the next 7 days, starting from our made-up day. We need to do this to be able to instantiate `ClientStub` so that it can respond according to these values.

From the name of the test, which should be structured as `Method_Condition_Expectation`, we can figure out what we are trying to do in this test. The weather that we get in reality contains 7 days starting from today, but what we return in `WeatherForecast[]` is the forecast from the next day onward for the next 5 days. Therefore, we ignore today's weather and use it nowhere.

The stub has shielded us from reaching to the real weather service and provided the canned values that we included in the `Arrange` section. If we were to call the real service, we would get unpredictable weather, from a test point of view, for unpredictable days (depending on when we run the test), which wouldn't make us able to write our `Assert` criteria.

This test is not enough to have good coverage of all the criteria that should be tested. You can find more tests for the `GetReal` method using the `ClientStub` class in the source code for this chapter in the `WeatherForecastControllerTests` class. The tests are:

```
GetReal_5DaysForecastStartingNextDay_
    WF5ThDayIsRealWeather6ThDay

GetReal_ForecastingFor5DaysOnly_WFHas5Days

GetReal_WFDoesntConsiderDecimal_
    RealWeatherTempRoundedProperly

GetReal_TodayWeatherAnd6DaysForecastReceived_
    RealDateMatchesNextDay

GetReal_TodayWeatherAnd6DaysForecastReceived_
    RealDateMatchesLastDay
```

I encourage you to have a look at the companion code to familiarize yourself with other examples.

Spies

Spies are extra functionality added to a stub class to reveal what happened inside the stub. For example, consider this business requirement where we need to ensure that we are only passing metric temperature (degrees Celsius) requests to *OpenWeather*.

We need to modify our stub to reveal what has been passed to OneCallAsync. The new code in the stub will look like this:

```
public Units? LastUnitSpy { get; set; }

public Task<OneCallResponse> OneCallAsync(decimal latitude,
    decimal longitude, IEnumerable<Excludes> excludes,
      Units unit)
{
    LastUnitSpy = unit;
    const int DAYS = 7;
    // the rest of the code did not change
```

We have added a property called LastUnitSpy to store the last requested unit and ended it with the Spy suffix as a convention. Our unit test will look like this:

```
public async Task
    GetReal_RequestsToOpenWeather_MetricUnitIsUsed()
{
    // Arrange
    var realWeatherTemps = new double[] { 1,2,3,4,5,6,7 };
    var clientStub = new ClientStub(
        default(DateTime), realWeatherTemps);
    var controller = new WeatherForecastController(null!,
        clientStub, null!, null!);

    // Act
    var _ = await controller.GetReal();

    // Assert
    Assert.NotNull(clientStub.LastUnitSpy);
    Assert.Equal(Units.Metric,
        clientStub.LastUnitSpy!.Value);
}
```

Notice that in this test, we did not populate the forecast temperature with meaningful values and used default `DateTime`. This emphasized to future test readers (other developers reading the code) that we don't care in this test for the variation of these parameters. We just wanted dummy objects to instantiate the `clientStub` object.

The last assert has validated receiving `Units.Metric`, which satisfies our business requirement.

You can add spies on-demand according to your tests and you can organize them in the way you like and hopefully, by now, the idea behind calling it *spy* makes sense.

Pros and cons of using stubs

Using stubs is simple and leads to readable code. Not needing to learn any special stubbing framework is also an advantage.

The problem with stubbing is that the more sophisticated your scenarios are, the more stub classes you need (`ClientStub2` and `ClientStub3`) or the cleverer your stub implementation needs to be. Your stub should contain minimal *cleverness* and business logic. In real-life scenarios, your stubs will be chunky and harder to maintain if you and your team are not careful about maintaining them.

Recap of the previous scenario

We have followed these steps to unit test the `GetReal()` method:

- We realized that `_client` is a dependency used by our SUT.
- We want to isolate our `GetReal` method from calling the real OpenWeather, so we need to provide an alternative behavior for `_client`.
- `_client` is an object of a class that implements the `IClient` interface.
- At runtime, the SUT is instantiated by the start up class. `Client`, which is provided by a third-party library, is passed to SUT. This `Client`, which implements `IClient`, provides a means for retrieving the real weather data from *OpenWeather*.
- Unit testing should not extend the test to a third party and should restrict the test to the SUT.
- To bypass calling the real service, we stubbed a class and called it `ClientStub`, and implemented `IClient`. `ClientStub` contains an implementation to generate made-up weather data.
- We wrote our unit test following the unit test name convention and the AAA structure.
- Our SUT constructor requires an instance of `IClient`, so we passed `ClientStub` to it.
- We can now test our SUT.

Stubs satisfy the first three test double requirements of those we stated previously. Also, with the help of spies, they satisfy the fourth requirement.

The same stubbing process is used for the rest of the unit tests for the GetReal method. Some teams use stubs as the main type of test doubles and other teams prefer to use mocks, which naturally leads us to our next topic.

Mocks

Mocks have great similarities with stubs, but rather than having the implementation of stubs done in regular coding, they use a *trick* to generate a behavior without having to implement a complete class. Mocks use third-party libraries to reduce the amount of coding involved to create a test double.

Mocking libraries

With mocks, you have to use a third-party library or build your own—heaven forbid. Two popular libraries for .NET are **Moq** (pronounced *mock you*) and **NSubstitute**.

- Moq started gaining popularity in 2010. It relies heavily on lambda expressions, which made it less verbose compared to its peers at that time. If you like lambda expressions, then Moq is for you.

- NSubstitute was also released near Moq time. Its focus is to provide a readable syntax for mocking.

Both libraries are mature in features and have a great online community. This book will use *NSubstitute*, but will also give you a quick introduction to Moq in the appendix.

To install NSubstitute, you can go to the unit test project directory and execute the following code:

```
dotnet add package NSubstitute
dotnet add package NSubstitute.Analyzers.CSharp
```

The second line is optional. It adds the C# NSubstitute analyzers, which uses Roslyn to add code analysis during compilation to detect possible errors. Also, it adds the capability for VS to give you hints to improve your mocks code.

You now have the NSubstitute library installed and ready to use.

Example using mocks

Mocks and stubs can be used interchangeably, so a good way of understanding them is to start from our previous stub implementation. Let's take the same example that we used in the stubs, which is testing GetReal. In that example, we used stubbing as our test double. Now, we use mocking, so we take the same test above and replace the Arrange part with this:

```
// Arrange
...
//var clientStub = new ClientStub(today, realWeatherTemps);
```

```
var clientMock = Substitute.For<IClient>();
clientMock.OneCallAsync(Arg.Any<decimal>(),
  Arg.Any<decimal>(), Arg.Any<IEnumerable<Excludes>>(),
  Arg.Any<Units>())
  .Returns(x =>
  {
      const int DAYS = 7;
      OneCallResponse res = new OneCallResponse();
      res.Daily = new Daily[DAYS];
      for (int i = 0; i < DAYS; i++)
      {
          res.Daily[i] = new Daily();
          res.Daily[i].Dt = today.AddDays(i);
          res.Daily[i].Temp = new Temp();
          res.Daily[i].Temp.Day =
             realWeatherTemps.ElementAt(i);
      }
      return Task.FromResult(res);
  });
var controller = new WeatherForecastController(null!,
  clientMock, null!, null!);
```

When using stubs, we coded an entire class so that we can instantiate it, as you can see in the commented line. In mocking the magical method from NSubstitute, `Substitute.For` created a concrete class from `IClient` and instantiated it all in one simple line.

However, the created object, `clientMock`, does not have any implementation for the `OneCallAsync`, so we have used NSubstitute methods to say: whichever parameter (`Is.Any<>`), that is passed to the `OneCallAsync` method, `Return` what is described in the provided lambda. The lambda content is the same content that was used in the `ClientStub` before.

We have dynamically attached a method implementation to an object that we just created with a couple of lines of code. This is pretty impressive and has less code than its previous stub counterpart. Mocking libraries have the ability to create a concrete implementation of an abstraction and, in advanced scenarios, they can mock concrete classes and substitute part of their implementation.

Of course, if you are using mocking, the `ClientStub` stubbing class that we used in the stubbing example is not needed. You only pick one or the other.

I have created a test class called `WeatherForecastControllerTestsWithMocking` to differentiate from the one that uses a stub. In a real-life project, you wouldn't do that as you will be typically using stubbing or mocking. This chapter and *Part 2, Building an Application with TDD*, will have dozens of examples that use mocks.

Spies

When it comes to mocking, we rarely use the term *spy* because the spy functionality is always embedded in the mocking framework. Spying in stubs is something you need to code, while for mocks, spying is built in. To illustrate this, it's best to take the spying-with-a-stub example that we presented earlier and make it spying with mock:

```
public async Task
    GetReal_RequestsToOpenWeather_MetricUnitIsUsed()
{
    // Arrange
    // Code is the same as in the previous test

    // Act
    var _ = await controller.GetReal();

    // Assert
    await clientMock.Received().OneCallAsync(
        Arg.Any<decimal>(), Arg.Any<decimal>(),
        Arg.Any<IEnumerable<Excludes>>(),
        Arg.Is<Units>(x => x == Units.Metric));
}
```

The `Arrange` and `Act` sections have not changed; we are only ignoring the output of the act stage. What has changed is our assertion. NSubstitute provides a method to spy on the passed parameters called `Received` and combines it with `Arg.Is` to verify what was passed.

This is the first example where the `Assert` section does not use xUnit's `Assert` class. This is perfectly legal, as the `Received()` method is an assertion method itself.

Pros and cons of using mocks

Mocks produce succinct code. They are slightly harder to read than plain code (code without a mocking library) if we compare them to stubs.

The drawback of mocks is that you depend on a library such as NSubstitute, and there is a learning curve associated with that. Also, some practitioners don't like the magic that mock libraries employ to attach behavior dynamically and prefer to keep things more obvious by using plain code (stubs).

Next, I summarize the differences between mocks and stubs.

Mocks versus stubs

The difference between mocks and stubs is important, as you need to be armed with logic to choose the best technique to suit you. Here is a quick list of differences:

- Both mocks and stubs are categorized as test doubles, and you can use one or the other in your project based on the project requirements or the team preference, although in the industry, mocks are used more often than stubs.

- Mocks are implemented with the help of a third-party library such as Moq or NSubstitute, while stubs do not rely on a library.

- Mocks are less verbose than stubs, but their syntax is slightly harder to read than plain code.

- Mocks are claimed to do some magic, which some practitioners believe corrupts the unit test, while stubs are plain code with no magic.

> **Important Note**
> The difference between mocks and stubs is a popular interview question. It is also important to expand on the answer and mention that both are test double types and are mainly used with unit testing.

Recap of the previous scenario

To recap, we had the same scenario as of the stub, but when stubbing, we have added a class to contain our stub and used it in the unit test. In mocking, we have used a mocking framework that facilitated including our implementation within the body of the unit test.

Mocks satisfy all test double requirements that we stated above. I hope the previous examples have given you a flavor of mocking. Next, we will explore another test double type.

Fakes

Fakes are libraries that mimic part or all of a real-life equivalent, and they exist in order to facilitate testing.

> **Important Note**
>
> The term *fake* has multiple definitions in the industry. This chapter is using the definition by Martin Fowler (`https://martinfowler.com/bliki/TestDouble.html`), as follows: "*Fake objects actually have working implementations, but usually take some shortcut which makes them not suitable for production (an InMemoryTestDatabase is a good example).*"
>
> One confusing name is a .NETframework called Microsoft Fakes that does isolation!
>
> One of the most popular examples library called **FakeItEasy** that does mocks. Also, Microsoft has a framework called Microsoft Fakes that does isolation!

One of the most popular examples of a fake within the .NET library is *Entity Framework Core In-Memory Database Provider*. This is a quote from the Microsoft documentation (`https://docs.microsoft.com/en-us/ef/core/providers/in-memory`):

This database provider allows Entity Framework Core to be used with an in-memory database. The in-memory database can be useful for testing, [...]. The in-memory database is designed for testing only.

When storing in memory, it is easy to wipe and recreate the storage while executing every individual unit test. That helps in repeating the tests without worrying about the changed state of the data. Although, if the storage was persisted on disk, say with a real database (SQL Server, Cosmos, Mongo or others), then resetting the data before every test is not a simple task. The volatile nature of an in-memory database is suitable for unit testing.

If *Test A* changed a username from JohnDoe to JohnSmith and *Test B* tried to change JohnDoe to JaneSmith, *Test B* will definitely fail if the changes that *Test A* had done were permanent (persisted to a physical disk database). Using a volatile in-memory database can make resetting the data easier in between every test. This is an important unit test principle known as **No Interdependency**.

Fakes are meant to help provide an implementation of a complex system to try to make your unit tests more realistic. If you have a system that uses a relational database and relies on EF Core, then the previous provider might help when unit testing:

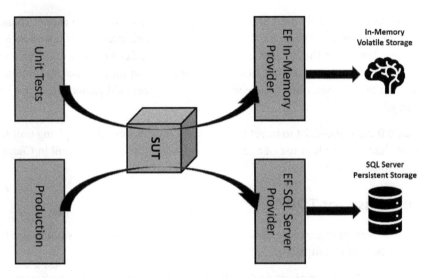

Figure 4.3 – In-memory storage versus production storage

Fakes satisfy the first three test double requirements of those we stated earlier. If fakes were to embed spying behavior, they would satisfy the fourth requirement.

In *Part 2, Building an Application with TDD*, we will be using this provider and we will show the usage of fakes.

Isolation

Isolation is not something you do with TDD, but I have added a limited introduction here for completeness. Isolation bypasses traditional DI all together and uses a different technique for DI known as **shim**. Shim involves modifying compiled code behavior at runtime for injection.

Owing to the complexity of the functionality of isolation frameworks, there aren't many of them in .NET. Here are possibly the only two frameworks available for .NET **Core**:

- **Microsoft Fakes**: Available with the Enterprise version of Visual Studio
- **Telerik JustMock**: A third-party commercial tool. It also has an open source restricted implementation called **JustMock Lite**.

I am unaware of a fully-featured isolation library that has a permissive, or free, license for .NET 5 and above.

Isolation frameworks are primarily used to unit test legacy systems, where you cannot change the production code to support DI. Therefore, the isolation framework injects the dependencies into the SUT at runtime. The reason they are not used with TDD is that TDD is about adding tests while gradually *modifying* production code, but isolation is not meant to modify the production code. Despite the fact you can use isolation frameworks to unit test greenfield projects, they are not the best tools for the job.

Although isolation frameworks exist to target legacy code, I don't believe that applying unit tests to code that cannot change is the best use of a team's time. I cover this in more detail in *Chapter 12, Dealing with Brownfield Projects*.

What should I use for TDD?

Let's start with the process of elimination. Isolation and isolation frameworks may not be used in the context of TDD as they are incompatible.

Dummies can coexist with all types of test doubles. Using a `NullLogger<>` service where the logger is not used or passing null arguments is going to happen in most of your unit tests. So, use a dummy when you can; in fact, using dummies should take precedence over other types if this is possible.

Teams usually use mocks or stubs, but not both unless the project is in a transitional state from one to the other. The debate of which one is better cannot be resolved in this book, it is all over the internet. However, given that stubs are harder to maintain and they need to be built up manually, mocks can do better as a start. Begin with mocks, get experienced, and then you can decide whether stubs can serve you better.

Finding appropriate fakes is a hit and miss. Sometimes, you can find a well-implemented fake such as the EF Core In-Memory Database Provider, and sometimes you might find an open source fake for some popular system. Sometimes, you may be unfortunate and have to create one yourself. But fakes are used in conjunction with mocks or stubs; as we will see in *Part 2* of the book, they are not one or the other. They add value to your tests, and you need to decide when to use them on a case-by-case basis.

In summary, for any object that should not be part of the SUT or for an unused dependency, use a dummy. To build and test dependencies, use mocks. Add fakes where it makes sense.

The meaning of stubs, mocks, and fakes will vary, and the definitions are muddled. I have tried to use the most common terminology in the industry. What is important is understanding all the test double options that we can use and using them appropriately.

Test doubles are what make unit testing different from other categories. This can be clarified further when we discuss other testing categories to better understand the uniqueness of unit testing.

More testing categories

You've probably heard about plenty of testing categories other than unit testing. There is **integration testing**, **regression testing**, **automation testing**, **load testing**, **pen testing**, **component testing**—and the list goes on. Well, we won't cover all these test categories, as explaining them all will not fit in this book. What we will do instead is discuss the two categories that have commonalities with unit testing. The first one is **integration testing** and the second one is what I call **Sintegration testing**. We will also have an honorable mention of **acceptance testing** due to its importance in building a full test categories suite.

Unit testing, integration testing, and Sintegration testing have one major difference that sets them apart. It is how they deal with dependencies. Understanding the differences will help clarify how unit testing fits into the test ecosystem.

Integration testing

Integration testing is, fortunately, easy to understand. It is exactly like a unit test, but with real dependencies, not with test doubles. An integration test executes an endpoint, such as a method or an API, which will trigger all the real dependencies, including external systems such as a DB and tests the outcome against a criterion.

Example

The xUnit framework can also execute integration and Sintegration tests, so to show an example, we can create an integration test project in the same way we created a unit test project. From your console, go to your solution directory and execute the following:

```
dotnet new xunit -o Uqs.Weather.Tests.Integration -f net6.0
dotnet sln add Uqs.Weather.Tests.Integration
```

We have just created an integration test project using the *xUnit framework* and added it to our solution. Our integration test will be going over HTTP and will be deserializing the JSON values, so we will need to do this:

```
cd Uqs.Weather.Tests.Integration
dotnet add package System.Net.Http.Json
```

This will add the .NET JSON NuGet package to your Integration test.

> **Important Note**
> Notice that the integration test project is not referencing Uqs.Weather. This is because the integration test project will trigger the RESTful API through HTTP and does not need to use any type from Uqs.Weather.

In this example, we want to test getting 5 days, starting from the next day:

```
private const string BASE_ADDRESS = "https://localhost:7218";
private const string API_URI = "/WeatherForecast/
    GetRealWeatherForecast";
private record WeatherForecast(DateTime Date,
    int TemperatureC, int TemperatureF, string? Summary);
```

On the class level, we add these fields. They specify the address of the service, which is pointing to my local machine, and also specify the URI of the SUT. By looking at the `WeatherForecast` class from `Uqs.Weather`, I know that I am getting back an array of `WeatherForecast` of five fields. So, I have constructed a similar record to the expected data that will come from the RESTful API call.

My integration test looks like this:

```
public async Task
    GetRealWeatherForecast_Execute_GetNext5Days()
{
    // Arrange
    HttpClient httpClient = new HttpClient
    { BaseAddress = new Uri(BASE_ADDRESS) };
    var today = DateTime.Now.Date;
    var next5Days = new[] { today.AddDays(1),
        today.AddDays(2), today.AddDays(3),
        today.AddDays(4), today.AddDays(5) };

    // Act
    var httpRes = await httpClient.GetAsync(API_URI);

    // Assert
    var wfs = await
    httpRes.Content.ReadFromJsonAsync<WeatherForecast[]>();
    for(int i = 0;i < 5;i++)
    {
        Assert.Equal(next5Days[i], wfs[i].Date.Date);
    }
}
```

We don't know on which day the test is going to execute, so we are fetching today's date and then calculating the dates of the next 5 days. We are creating and setting up an `HttpClient` to issue an HTTP call.

In `Act`, we are calling the RESTful API's endpoint.

In `Assert`, we are converting the returned value from JSON to the `record` class that we created earlier and checking that we are getting the next 5 days.

This test requires a different setup to run than the way we were running unit tests before. This is an out-of-process test, which means that the API is running on one process and the test is running on another process. The two processes are communicating with each other via HTTP. So, to run this test, we first need to initiate the REST API process. Right-click on **Uqs.Weather** | **Debug** | **Start Without Debugging**. Then this will launch Kestrel web server in a console window and make our API ready for an HTTP call.

Now, you can go ahead and execute the integration test in a similar fashion to executing unit tests.

Activities triggered by this test

The API call that we have just executed in our integration test triggered multiple dependencies to generate the output. These are a few of the triggered dependencies:

- The network, including the HTTPS connection, between the integration test and the ASP.NET Web API host
- The ASP.NET Web API host, which spun a process and displayed in a console window
- The routing code that analyzed the request and spun the right action method in the controller
- The DI container that decided which object to create and inject
- The HTTPS connection between `Uqs.Weather` and the *OpenWeather* API

We know that each of these dependencies works on its own. By executing this test, we made sure that all our components are integrated and work well together. The following diagram shows a hint of what happened when we executed the test:

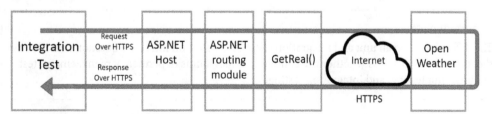

Figure 4.4 – Request/response going through the dependencies

These are not all the components, as I have missed some, but hopefully, you've got the idea.

Points to note

This is our first integration test in this book (and the last one), and I would like to point out the differences between this test and its unit test equivalent:

- We do not know the day in advance—we have to determine the date dynamically, while in our unit test, we had a pre-configured day in the `Assert` section.

- There are no test doubles, and everything is executing the real objects.

- We had two processes running; one is the web server that hosts the API, and the other is our integration test. Unit tests were calling directly, in-process.

- There is a probability of failure that is not related to the test. The test could fail due to:

  - Firewall issues

  - The *OpenWeather* service is down

  - Exceeding the number of calls permitted by our *OpenWeather* license

  - ASP.NET Web server has not started

  - The routing module is not configured properly

  - Other environmental issues

- This test, while not measured, will take a longer time as it is communicating between two processes, hitting multiple components, and then going over HTTPS, which includes serialization/ deserialization and encryption/decryption. Although the execution time is unnoticeable, having 10 seconds of integration tests will add up.

Unit testing versus integration testing

Integration tests and unit tests are great tools, and comparing them might imply using one over the other. This is not the case as both complement each other. Integration testing is good at testing a full cycle call, while unit testing is good at testing various scenarios of business logic. They have separate roles in quality assurance; the problem arises if they step on each other's toes.

> **Important Note**
> The difference between unit and integration testing is a common interview question that allows the interviewer to assess whether the candidate understands dependency management, test doubles, unit testing, and integration testing.

Advantages of integration tests over unit tests

Here are the advantages that might lead us to use integration tests over unit tests:

- Integration tests check the real behavior that mimics what the end client might do, while unit tests check what the developer thinks the system should do.

- Integration tests are easier to write and understand as they are regular code that doesn't use test doubles and doesn't care about DI.

- Integration tests can cover scenarios that unit tests cannot cover efficiently, such as the integration between the components of the whole system and DI container registration.

- Integration tests can be applied to legacy systems or greenfield systems. In fact, integration tests are one of the recommended ways to test legacy systems, while unit tests require code refactoring to be introduced into brownfield projects.

- Some integration tests, like the example above, can be written in a language-agnostic way. So, the previous test could have been written in F#, Java, or Python, or by a tool such as Postman, while unit tests use the same language as the production code (C#, in our case).

Advantages of unit tests over integration tests

Here are some advantages that might lead us to use unit over integration tests:

- Unit tests are much faster to execute, and this is very important when running hundreds of tests and looking for a short feedback loop, especially before integrating the code or releasing to an environment.

- Unit tests have predictable results and are not affected by time, third-party service availability, or environmental intermittent issues.

- Unit tests are repeatable as they do not persist any data, while integration tests might change the data permanently, which may make subsequent tests unreliable. This happens when writing and editing. Our example above was reading (Getting), so it did not suffer from this problem.

- Unit tests are easier to deploy to CI/CD pipelines (we will demonstrate this in *Chapter 11, Implementing Continuous Integration with GitHub Actions*).

- Spotting bugs in unit tests is done sooner and can be pinpointed faster than finding the same bug in integration tests.

- Unit tests can run during feature development, while an integration test can only be added when a feature is fully ready.

Confusing unit and integration tests

Frameworks such as xUnit or NUnit are used in various integration test implementations. The term *Unit* in the framework names might mislead some developers to think what is written in these projects is unit tests. Add to that using the *AAA convention* and the method name convention, and this might mislead as well. In fact, I used these conventions in the previous integration test, but using the same convention of a unit test does not make a test a unit test.

Setting up the infrastructure and building the CI pipeline will vary based on the type of test being implemented. Although they look the same, it is important to differentiate between them to understand the level of tasks and maintenance required.

Given that they look the same, how do you spot one or the other? There is a telltale, if they are not relying on the real dependency, they are most likely to be unit tests. If they are using real objects that would trigger real external dependencies, chances are they are not unit tests.

Sintegration testing

A **Sintegration** test is a midway between integration testing and unit testing. Integration testing relies on real components, while unit testing relies on test doubles. Sintegration testing tries to solve the shortcomings of integration testing by mixing elements of unit testing:

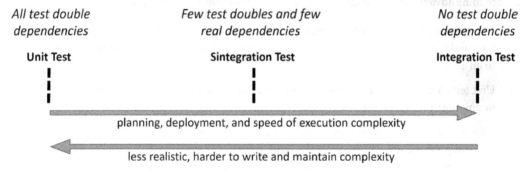

Figure 4.5 – Unit, Sintegration, and integration testing

I have found some developers referring to this type of test as a **component test**. But a component test, in software engineering, means something different, and I feel the developers were—rightly—more concerned about what it does than naming it correctly. This test category has these distinctive features, as outlined here:

- Substitutes (swaps) some dependencies with their real counterpart
- Simulates some real dependencies by building fakes (the fakes of test doubles)

For the *S* at the beginning of *substitute, swap,* and *simulate,* and for its resemblance to integration tests, I gave it the name *Sintegration testing.* As always, let me clarify Sintegration tests with an example.

Example

Let's assume that we have a web project that uses logging, service bus queues, and Cosmos DB. The logging logs to the cloud, so it needs cloud connectivity. The queue is also a cloud component, and Cosmos DB is a cloud component as well.

Let's also assume that we have a series of APIs to deal with the user profile, such as the `UpdateName` API and the `ChangePassword` API. A Sintegration test can do the following:

- Use the Kestrel web server and have the same features as per production, as Kestrel is flexible enough to run on-demand on local machines, test, and production environments.

- Writing logs will require access to the cloud, so we inject a `NullLogger<>` service that will ignore logging but will let the system work.

- Queues are only available on the cloud, so we replaced this with a fake in-memory queue that can be easily wiped between tests.

- Cosmos DB does not have an in-memory implementation, but the cloud version can be easily wiped between test runs. So, we use the same .NET Cosmos DB client library, but we point to a different database—the Sintegration Tests Cosmos DB, at test time.

Here is a project component diagram of what the system would look like:

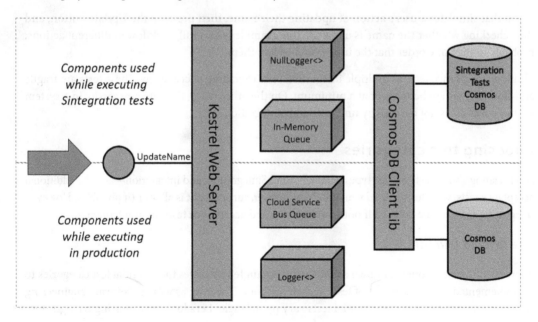

Figure 4.6 – Executing Sintegration tests

The Sintegration tests in this scenario use fakes, dummies, and real components. They cover part of the integration in the system, on the other hand, the queues and DB can be easily wiped between individual tests to make sure no Sintegration test is affecting other tests.

Sintegration testing started gaining traction in recent years; perhaps the movement from the classical .NET Framework to .NET has surfaced this type of testing as .NET Core libraries are no longer dependent on specific Windows components, such as the IIS web server. Also, ASP.NET Core has specific implementations that allowed this type of testing, whereas, in the past, these implementations were not part of the framework. One of these implementations is having a Kestrel web server as part of .NET and being able to spin it easily without relying on a special deployment.

This was a quick overview of Sintegration tests. While Sintegration tests are not part of TDD, they are important to understand as they are related to unit tests.

Acceptance tests

Acceptance testing is the testing of a complete functionality or a feature end-to-end. One popular tool for acceptance tests that is used for websites is **Selenium**. You might find this category of testing under different names, such as **functional testing**, **system testing**, and **automated testing**.

Example

A test that simulates a user's action of updating their name, then pressing the **Update** button, and then checking whether the name is updated. This example tests a full workflow of different actions, probably in the same order that the user would trigger them.

Think of acceptance tests as multiple integration tests executing successively. These tests are fragile and slow, so they are better kept at a minimum. On the other hand, they are a necessity in a system as they cover areas not covered by unit tests and Sintegration tests.

Choosing test categories

After reading about the three test types—namely, unit, Sintegration, and integration—and the additional acceptance tests, what tests should you write? The answer, surprisingly, is all four (if possible). However, integration tests can be omitted if both Sintegration and acceptance tests are implemented.

The testing triangle

There is an industry concept known as the **testing triangle** that states the essential test categories to be implemented and the number of tests in each category. However, as with all software engineering concepts, the categories of the essential tests vary per triangle. Let's look at the testing triangle for a greenfield project:

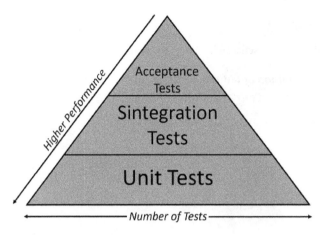

Figure 4.7 – Testing triangle

The triangle in the preceding image advocates more unit tests than the other two testing categories. Assume you are experienced with the three mentioned tests. You will find that implementing unit tests takes the least amount of time compared to the other two and takes a few seconds to execute hundreds of them (after compiling and loading the code).

Discussing this triangle with respect to brownfield projects will be covered in *Chapter 12, Dealing with Brownfield Projects*. Hopefully, debating integration test and Sintegration test have helped you to better understand what a unit test is.

Summary

In this chapter, we compared unit tests with its siblings: *integration* and *Sintegration tests*. We listed test doubles and gave an example of each, and we have also seen xUnit and NSubstitute in action.

Our journey with understanding unit testing and test doubles will not stop here, but we will cover more examples of the two topics across the rest of the book.

So far, you can consider the experience from this chapter to take you to TDD level 3 out of 5! And now, you should be able to write a basic unit test that uses test doubles.

We have not covered the advantages and disadvantages of unit testing—yes, it has disadvantages! We have also not covered how TDD relates to unit testing and the best practices of unit testing because this is the role of the next chapter, *Test-Driven Development Explained*.

Further reading

To learn more about the topics discussed in the chapter, you can refer to the following links:

- *Martin Fowler's definition of test doubles*: `https://martinfowler.com/bliki/TestDouble.html`

- *NSubstitute*: `https://nsubstitute.github.io`

Test-Driven Development Explained

Test-Driven Development (TDD) is a set of practices on top of unit tests. They alter the way you design your code and you write unit tests. Basically, it is a different approach to writing code than the classical technique of writing code and then testing it after.

It is a cliché to say that TDD is not just doing testing first, but rather than me telling you otherwise, you will decide yourself after going through *Chapters 5* and *6*.

In this chapter, we will:

- Go through TDD pillars
- Implement a software feature following the TDD style
- Converse the FAQs and criticism around the subject
- Discuss having TDD with Sintegration testing

By the end of the chapter, you will be able to use TDD to write basic coding tasks and understand the topics around the subject and where TDD fits in the software ecosystem.

Technical requirements

The code for this chapter can be found at the following GitHub repository:

```
https://github.com/PacktPublishing/Pragmatic-Test-Driven-Development-
in-C-Sharp-and-.NET/tree/main/ch05
```

TDD pillars

TDD is a set of practices that specify how and when a unit test should be written. You can write unit tests without TDD, but TDD has to have a type of test associated with it. Sometimes, you can hear TDD and unit tests used as if they mean the same thing, but they are not.

While the ecosystem around TDD is sophisticated as it touches a lot of software engineering aspects, TDD as a standalone concept is easy to explain and understand. We can summarize TDD as these two pillars:

- **Test first**
- **Red, Green, Refactor (RGR)**

Let's discuss these pillars.

Test first

The idea here is to start with the tests before starting with the production code. It really means testing code that doesn't exist yet!

Testing first changes the way we write code, as now you are redirected to think about your classes' structure and public methods before the implementation. This encourages the developer to reflect on the design from the client's perspective (the client is the external code calling your code, also known as the caller).

We will demonstrate how to start with tests by going through a few examples to get you accustomed to the concept and familiarize you with how it is done. Then, we will go into detail about the benefits of this approach in the first section of *Chapter 6, The FIRSTHAND Guidelines of TDD*.

Red, green, refactor

The RGR process is the one used when writing code in TDD-style. This process works in tandem with the test first guideline. This is how it goes:

1. You are planning to write a coding task, which is part of a feature that you want to add to the code.

2. You write the unit test for this production code while the production code doesn't exist (as you haven't written it yet). Also, maybe you are planning to update existing production code, so you write a unit test first, assuming the final production code is in place.

3. You run the unit test and it will fail (red), for one of two reasons:

   - The code doesn't compile, as the production code is not written yet.

   - The test will fail, as while it is compiling, the logic that implements the new coding task is not the right one, as the existing production code has not been updated to reflect the new feature yet.

4. Write the fastest and minimal code to make the test pass (green). Don't perfect the code at this stage; you can also copy the intended code from the internet. The idea is just to get on with it.

5. Now, we know that our coding task is in place and it is working; however, you might consider refactoring the production code if it suffers from one of these issues:

    - Readability

    - Performance

    - A design that doesn't comply with the rest of the code

6. Run your unit test to make sure you have not broken anything while you were refactoring. If it is broken (red), then you would naturally go back to *step 3*.

You can see why the color names were picked:

- **Red**: Represents failed tests, which look red in the test runners (such as VS Test Explorer)

- **Green**: Represents passed tests, which look green in the test runners

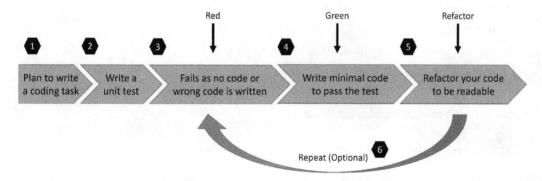

Figure 5.1 – The red, green, refactor process

The previous diagram highlights the steps that we've just discussed in the RGR process.

We can draw diagrams and talk more about TDD, or we can demonstrate with examples, which is what we will do next.

TDD by example

TDD is best understood by looking at examples, so let's take a story and write it TDD-style. We will be working on the feature described by this story:

Story Title:

Changing a username

Story Description:

As a customer

Given I already have an account

When I navigate to my profile page

Then I can update my username

Acceptance Criteria:

The username can only contain between 8 and 12 characters, inclusive:

- Valid: AnameOf8, NameOfChar12

- Invalid: AnameOfChar13, NameOf7

Only alphanumerics and underscores are allowed:

- Valid: Letter_123

- Invalid: !The_Start, InThe@Middle, WithDollar$, Space 123

If the username already exists, generate an error

Let's not waste any time and implement this story.

Creating the solution shell

We will create a class library called `Uqs.Customer`, add a unit testing project to test it called `Uqs.Customer.Tests.Unit`, and add them to a solution called `TddByExample.sln`. So, let's start:

1. In a directory called `TddByExample`, create the `lib` class and the following xUnit projects:

   ```
   dotnet new classlib -o Uqs.Customer -f net6.0
   dotnet new xunit -o Uqs.Customer.Tests.Unit -f net6.0
   ```

2. Create a solution file and add the projects to it. The solution name will be the directory name. So, in this case, it will automatically be called `TddByExample.sln`:

   ```
   dotnet new sln
   dotnet sln add Uqs.Customer
   dotnet sln add Uqs.Customer.Tests.Unit
   ```

3. Reference the production code project from the unit tests project:

```
dotnet add Uqs.Customer.Tests.Unit reference Uqs.Customer
```

4. When you open the solution, you will get the following:

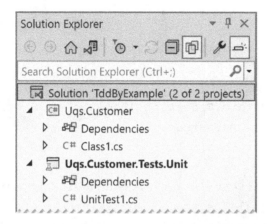

Figure 5.2 – Solution Explorer showing the newly created project

Now, to the coding stage.

Adding coding tasks

The requested feature comprises smaller coding tasks (challenges), so you will be adding multiple unit tests with at least one unit test for each coding challenge.

I start with the challenge that gives the shortest feedback and work my way up. This will help me build the initial structure without having to worry about everything at the same time. It will also help me avoid tasks such as connectivity to the database and availability of the username in the database (username is already taken) from the start. Then, start adding more sophisticated challenges in order of complexity. Let's see how we can achieve this next.

Coding task one – validate null username

I normally spend little time analyzing the shortest code challenge and use my gut feeling. I feel the shortest route is to check the username for nullability. Now, rename the template test class from UnitTest1.cs to ProfileServiceTests.cs and replace the content with this code:

```
namespace Uqs.Customer.Tests.Unit;


public class ProfileServiceTests
```

```
{
    [Fact]
    public void
       ChangeUsername_NullUsername_ArgumentNullException()
    {
        // Arrange
        var sut = new ProfileService();

        // Act
        var e = Record.Exception(() =>
            sut.ChangeUsername(null!));

        // Assert
        var ex = Assert.IsType<ArgumentNullException>(e);
        Assert.Equal("username", ex.ParamName);
        Assert.StartsWith("Null", ex.Message);
    }
}
```

The previous code does the preparation to pass `null` to our intended method. It executes the method and records `Exception` generated by the method.

In the end, we check whether we got an exception of the `ArgumentNullException` type having `username` as the argument and having the message starting with `Null`.

Before executing this, let's reflect on what has happened so far.

Naming your test class

Right off the bat, the decision that I had to take was to think of my production code class name, because I needed it so I could append the `Tests` suffix to it and create my test class name (remember the convention: `ProductionCodeClassTests`). I am following **Domain-Driven Design (DDD)** in my architecture, so I directly thought of this as a service class. Don't worry about DDD terminology and how it fits with TDD for now, as DDD will have a dedicated chapter, *Chapter 7, A Pragmatic View of Domain-Driven Design*.

I didn't give naming my class deep thought as I can always rename my classes with minimal effort. I landed on `ProfileService`, so I can call my unit test class `ProfileServiceTests`. Please note that I have not created `ProfileService` at this stage.

> **Note**
> Some developers write meaningless test class names and then rename them after finishing the first unit test. Do what makes you more productive. This is not a rigid process.

Unit test method naming

The moment I wanted to write my test, I needed to think about what I am testing and what I should expect, following the `MethodName_Condition_Expectation` approach. So, I chose my method name to be `ChangeUsername` and the condition to check if `null`.

Deciding your expectations upfront

The expectation requires a bit of a pause and think. I am expecting whoever calls this method to have not sent `null`, as they have checked it on the UI or by any other means. So, I will be merciless and throw an exception if it doesn't match my expectation and let the client deal with it.

The point here is that I directly thought from the client's perspective and focused on the external behavior of my intended method.

Making sure the test fails

At this stage, you can see VS highlighting your code with squiggly lines, so it doesn't take a genius to conclude that the code will not compile, as there is no production code written yet.

You have taken your first step in RGR, as you've got the red.

Creating the production code class shell

Let's at least be able to compile to get some IntelliSense help from VS. So, in your `Uqs.Customer` project, modify `Class1.cs` to be `ProfileService.cs`, and the content will be something like this:

```
namespace Uqs.Customer;
public class ProfileService
{
}
```

However, given that you will be doing this a lot, there are shortcuts for this. For example, if you have ReSharper, then it will give you the option to generate the production code based on your unit test. With VS, rather than renaming `Class1.cs` as in the previous step, just delete it, and follow these steps to do the same thing:

- Hover over `ProfileService()` where you have the squiggly line, and VS will show you a bulb. Expand the bulb and select **Generate new type...** as in the following figure:

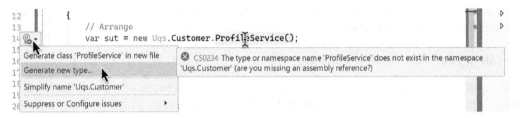

Figure 5.3 – Selecting the refactoring bulb

Note

Sometimes the bulb takes time to appear; you can always use the *Ctrl + .* shortcut to force it to appear.

- When the dialog box appears, you can change its settings as follows:

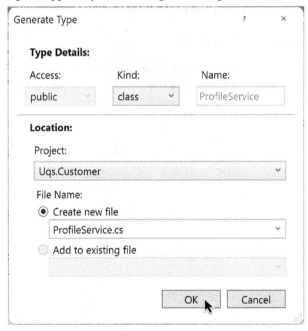

Figure 5.4 – Generate Type dialog box

This way, you will get the same `ProfileService.cs` class generated as before.

> **Note**
>
> Selecting the first option from the bulb menu, **Generate class 'ProfileService' in a new file**, doesn't do the job, as VS will generate the file in the unit testing project, while you intend to generate it in the production code project.

Now that we have our class shell created, let's continue our production code writing process.

Creating the production code method shell

To create the `ChangeUsername` method, hover over it, and select the bulb as follows:

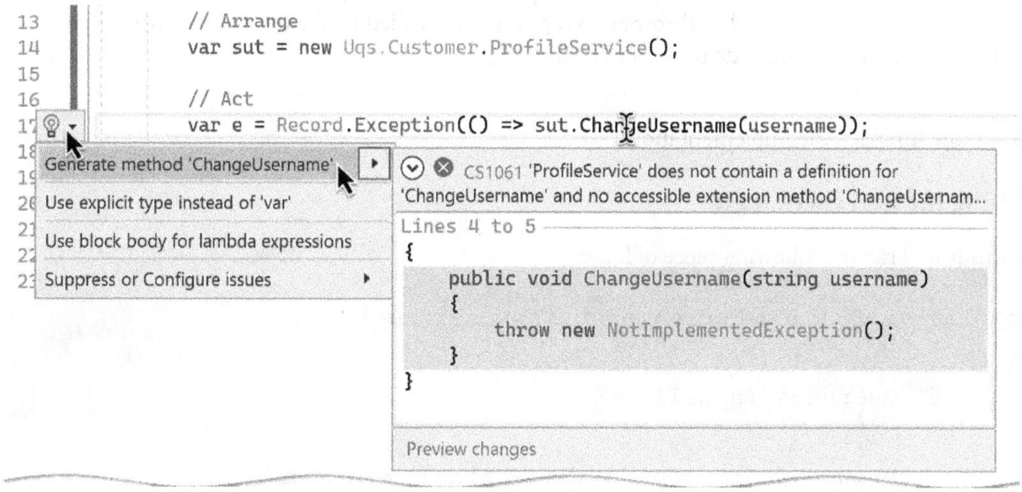

Figure 5.5 – Generate method

It will show you what is going to be generated in the display window. It is exactly what we want, so select **Generate method 'ChangeUsername'**. This will add this code to the `ProfileService` class:

```
public void ChangeUsername(string username)
{
    throw new NotImplementedException();
}
```

This is the generated code. Alternatively, you can write the code yourself.

> **Note**
>
> This is using C# 10, so not having `string?` (with a question mark after *string*) as the parameter warns the caller that this method doesn't expect `null`. But, the caller can still force a `null`. Notice that the unit test did force a `null` in the `Act` section by having a bang sign after the null: `sut.ChangeUsername(null!)`.

Notice that the generated code has already added for you a `NotImplementedException`.

> **Note**
>
> Having `NotImplementedException` is a good practice to highlight to the reader that the code is not yet written, and to throw an exception when called by mistake in case you forgot about it and pushed the code to source control.

Now to the fun part, the implementation.

Writing the null check logic

All this is to write the following piece of logic:

```
public void ChangeUsername(string username)
{
    if (username is null)
    {
        throw new ArgumentNullException("username", "Null");
    }
}
```

The first argument in the exception represents the parameter name, and the second represents the error message.

Run the unit tests from Test Explorer (*Ctrl + R, A*) and you should get the code compiling and all the tests passing (green).

Refactoring

I looked at the code after I'd written it and I thought it could be improved with the following:

Using a magic string inside the code to match my parameter name is not a good practice, as whenever I change the parameter name, the string doesn't necessarily change with it. I will use the `nameof` keyword.

My refactored code would look as follows:

```
throw new ArgumentNullException(nameof(username), "Null");
```

Now that I did these changes, I ran the tests again, and they passed.

While the size of the refactoring is small, and usually refactoring will be more sophisticated in a more streamlined example, this example serves as a good way to show you how you can implement an optional refactoring.

We have finished our first coding task! The first task is usually longer than the others as, in the first one, you will be building the shell and deciding about a few names. Our second task is going to be shorter.

Coding task two – validate min and max lengths

Again, without spending too much time thinking about the second thing to test, I thought of the length validation, according to the story where the length of the username should be between 8 and 12 characters included, so this is my second unit test to target this scenario:

```
[Theory]
[InlineData("AnameOf8", true)]
[InlineData("NameOfChar12", true)]
[InlineData("AnameOfChar13", false)]
[InlineData("NameOf7", false)]
[InlineData("", false)]
public void ChangeUsername_VariousLengthUsernames_
    ArgumentOutOfRangeExceptionIfInvalid
    (string username, bool isValid)
{
    // Arrange
    var sut = new ProfileService();

    // Act
    var e = Record.Exception(() =>
        sut.ChangeUsernam(username));

    // Assert
    if (isValid)
    {
        Assert.NullI;
```

```
    }
    else
    {
        var ex =
        Assert.IsType<ArgumentOutOfRangeException>(e);
        Assert.Equal("username", ex.ParamName);
        Assert.StartsWith("Length", ex.Message);
    }
}
```

The previous code prepares many testing scenarios for valid and invalid-length usernames. It uses the Theory attribute to pass multiple scenarios to the unit test. In the end, we check if we got an exception of the ArgumentOutOfRangeException type. We branch with an if statement, as a valid username does not produce an exception, so we will get null.

> **Note**
>
> Some practitioners are against having any logic, such as an if statement, within a unit test. I am with the other school that having light logic that is clear and readable would reduce the amount of repetition in unit tests. Do what feels readable for you and your team.

This sample test data may come from who wrote the story (for example, the product owner, business analyst, or product manager), from you, or a combination of both.

The red stage

Run the unit tests, and this newly added unit test should fail as we have no implementation written yet.

The green stage

Add this logic to your method:

```
if (username.Length < 8 || username.Length > 12)
{
    throw new ArgumentOutOfRangeException
        ("username","Length");
}
```

Run the unit tests from Test Explorer (*Ctrl + R, A*) and you should get the code compiling and all the tests passing. We've reached green.

The refactoring stage

I looked at the code after I'd written it and I thought it could be improved with the following:

> I am getting the length twice to compare it. Lucky for me, C# 8 introduced pattern matching that will lead to a more readable syntax (arguably). Also, C# might do some optimization magic to prevent the `Length` property from being executed twice.

My refactored code would look as follows:

```
if (username.Length is < 8 or > 12)
{
  throw new ArgumentOutOfRangeException(
    nameof(username), "Length");
}
```

Now that I did these changes, I ran the tests again, and they passed.

Coding task three – ensure alphanumerics and underscores only

As per the requirements, we are only allowing alphanumerics and underscores, so let's write the test for this:

```
[Theory]
[InlineData("Letter_123", true)]
[InlineData("!The_Start", false)]
[InlineData("InThe@Middle", false)]
[InlineData("WithDollar$", false)]
[InlineData("Space 123", false)]
public void
    ChangeUsername_InvalidCharValidation_
        ArgumentOutOfRangeException
        (string username, bool isValid)
{
    // Arrange
    var sut = new ProfileService();

    // Act
    var e = Record.Exception(() =>
        sut.ChangeUsername(username));
```

```
    // Assert
    if (isValid)
    {
        Assert.Null(e);
    }
    else
    {
        var ex =
            Assert.IsType<ArgumentOutOfRangeException>(e);
        Assert.Equal("username", ex.ParamName);
        Assert.StartsWith("InvalidChar", ex.Message);
    }
}
```

Run the test and it should fail except for the first test of `Letter_123`, which is a valid test. We want everything to fail to make sure we did not make a mistake. This is the output of Test Explorer:

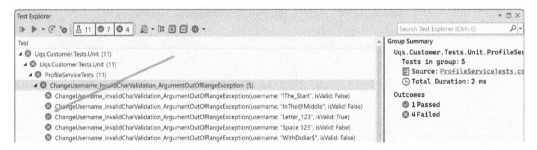

Figure 5.6 – Test Explorer output for the valid letters test

You can follow one of these two solutions to make the test fail::

1. Go to the production code and write code that will make this test fail. I don't personally like this approach, as I feel it is a purist approach, but there is nothing wrong with it.

2. Debug the code and see why it is passing without implementation. This is the approach that I took and it looks like this is a valid scenario anyway, so it should have passed. I can ignore the passed test and assume everything has failed.

So, let's write the proper implementation.

> **Note**
>
> You can see in all our tests that we are not asserting if there is an exception only. We are asserting the type of exception and two fields on the exception. This approach will help us catch the specific exception we are looking for and avoid catching other exceptions caused by something else.

The quickest way is to use a *regex* that will only allow alphanumerics and underscores. The fastest way is to search online for `alphanumeric and underscore only C# regex`. I found the regex on StackOverflow and it looks like this: `^[a-zA-Z0-9_]+$`.

Remember that my intention is to make this pass as soon as possible without thinking too much about the code or polishing it. This is the new code:

```
if (!Regex.Match(username, @"^[a-zA-Z0-9_]+$").Success)
{
    throw new ArgumentOutOfRangeException(nameof(username),
        "InvalidChar");
}
```

Run the test again and it will pass.

However, the code suffers from a performance problem, as having an inline regex is slow. Let me optimize the performance and improve the readability. This is my whole class after refactoring:

```
using System.Text.RegularExpressions;
namespace Uqs.Customer;
public class ProfileService
{
    private const string ALPHANUMERIC_UNDERSCORE_REGEX =
        @"^[a-zA-Z0-9_]+$";
    private static readonly Regex _formatRegex = new
    (ALPHANUMERIC_UNDERSCORE_REGEX, RegexOptions.Compiled);

    public void ChangeUsername(string username)
    {
        if (username is null)
        {
            throw new ArgumentNullException(nameof(username),
            "Null");
```

```
      }
      if (username.Length is < 8 or > 12)
      {

        throw new ArgumentOutOfRangeException(
          nameof(username), "Length");
      }
      if (!_formatRegex.Match(username).Success)
      {

        throw new ArgumentOutOfRangeException(
          nameof(username), "InvalidChar");
      }
    }
  }
```

Run again after refactoring. In all honesty, the test failed for me as I missed copying a letter from the regex while refactoring. Running the test again showed me that my refactoring wasn't right, so I fixed the code and tried again.

Coding task four – checking if the username is already in use

Obviously, checking if a username is already in use will require a trip to the database, and testing for that will require test doubles. Also, as you are doing IO operation (by accessing the DB), all your methods will be following the `async await` pattern.

This coding task and other coding tasks that are left for this feature to be completed will require database access, and I deliberately avoided this. I wanted this chapter to familiarize you with TDD without going through test doubles and more advanced topics. In *Part 2, Building an Application with TDD*, of this book, you will have dedicated chapters that will involve a mix of TDD, DDD, test doubles, and DBs. So, for now, we are going to stop here. Otherwise, how can I encourage you to continue reading if I explain everything here?

This ends our coding tasks for this chapter, and I hope so far you've got the TDD rhythm.

Recap

When starting a new feature, you need to think of this feature as a series of coding challenges. Every coding challenge will start with a unit test, similar to the following figure:

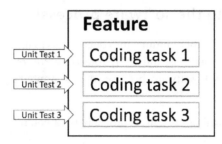

Figure 5.7 – Feature made of tasks, with a unit test targeting each task

Sometimes, you don't have a unit test project created yet, so you've seen how to create one. Sometimes, you add your unit tests to an existing unit testing project and you can immediately go ahead and start adding your tests.

You had to think upfront about how your client will interact with your production code and you designed everything according to the expectation of the client.

You followed the RGR pattern in adding each coding task and you've seen multiple examples of doing it.

More advanced scenarios are kept for *Part 2, Building an Application with TDD.*

FAQs and criticism

TDD is one of the most controversial topics in modern software development. You will find some developers swearing by it and others claiming it is useless.

I will try to objectively answer the questions and show you both views where relevant.

Why do I need to do TDD? Can't I just do unit testing?

As you've gathered from the start of the chapter, TDD is a style of writing unit tests. So, yes, you can write unit tests without following the TDD style. In the next chapter, you will find the *first guideline* from the FIRSTHAND guidelines that will focus on the benefits of following the TDD style.

I found that some teams are reluctant to do TDD for various reasons. My recommendation is not to abandon unit tests in case your team is not inclined to follow TDD. Maybe if you start with unit testing, then the next evolution would be TDD. This would slow down the rate of change for some teams.

Did I say this before? Don't abandon unit testing even if you are not following TDD.

TDD feels unnatural to the software process!

I trust that when you first learned programming, the concern was to understand basic programming constructs, such as `for` loops, functions, and OOP. Your concerns, or your tutor's concerns, were not producing scalable high-quality software, because you just wanted a program that worked, with bugs here and there.

That worked while you were learning, and this is probably what you refer to as *natural* because you have done this from day one.

In the real world, the modern expectations from software are to be:

- **Scalable**: Cloud-based solutions became the norm and microservices took control.
- **Automated**: The manual testing process turned out to be old-fashioned. A *developer in test* turned into a popular job title, and automated testing became the modern trend.
- **DDD**: Having objects interact with each other in a sophisticated manner.
- **Release pipeline ready (CI/CD)**: CI has a dedicated chapter in this book. In brief, your software should allow incremental feature addition and push to production every so often.

The preceding scenarios are the concerns of today's software developers and for that, you have to change your work strategy to the new real-life norm. This requires a paradigm shift in the way you write code, hence, the TDD norm of development.

Doing TDD is going to slow us down!

When starting a project, not doing any sort of testing would lead to faster results in the short term. Consider the following diagram that depicts this concept:

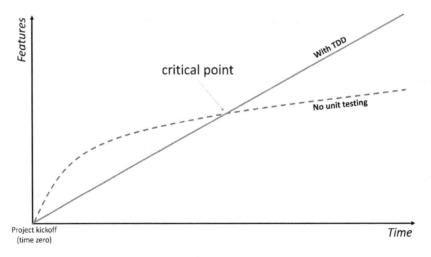

Figure 5.8 – TDD versus no testing concerning time and features

If you are starting from scratch building software with no dependencies and no features, it is easy to run doing no tests, maybe even having someone manually doing them. The project is small, manageable, and easy to change and deploy.

Test-free software will be fast to develop at the early stages until you start having a bunch of dependent features, where you change one and the other is broken! This is where test-backed software begins to shine, where adding a new feature has less probability of introducing a bug to an existing feature, as this should be caught by well-implemented tests. I have heard this phrase many times from product owners: *when we add a new feature, another unrelated area breaks!* They usually blame the developers or the tester who let this error pass. It is easy to think, for non-developers, that features are not related, but you and I know that this is not the case.

Obviously, after the *critical point*, the slowness in speed for non-tested software is due to the fact that the features are no longer manageable. The features are changing rapidly and the developers are moving to different areas or even leaving the project, and new developers are replacing them.

So, yes, TDD might slow you down, but this depends on where you are in the development process. If you passed the critical point in the preceding diagram, then you will start reaping the benefits of your investment.

Is TDD relevant for a start-up?

Developers involved in start-ups are usually stressed out and bombarded with feature requests. The company's survival and funding might depend on the next set of features. Rarely the product owners are concerned about the long-term strategy, because why bother if the destiny of the start-up is risky? We will worry about tomorrow later.

If the start-up may fail before the critical point (in the preceding chart), then would it be worth investing in TDD? But what if they passed this critical point and have no unit tests? Maybe the company will rewrite their codebase when they have financial backers and customers. They will add unit tests, or maybe they won't and they will struggle to add new features.

The start-up situation is complex; you can see this from the previous argument. The answer to this question depends on individual cases.

I don't like to do TDD, I like to do my code design first!

If you create your classes' structure before writing your tests, *you are still doing TDD*. Remember that TDD is a set of best practices and having your own flavor doesn't exclude you.

Unit testing doesn't test the real thing!

This is a criticism that the TDD community seeks to actively improve. This is mainly related to the usage of test doubles.

Test doubles try to mimic the behavior of real objects, and the problem here is in the word *try*. The problem of mimicking a real object is that it relies on the developer's best attempt to predict what the behavior of the real object is. This can be done in three ways:

1. Reading the documentation of the real object and trying to code something similar in a test double

2. Reading the source code, if available, and extracting its essence to build the test double

3. Doing a proof-of-concept sample to call the real object to check its behavior

These ways take research and experience. Sometimes, the test double object doesn't reflect the real object and that may lead to a wrong test and potentially to a bug. Let's take this example of a third-party method:

```
public string LoadTextFile(string path) {...}
```

The method above loads a text file and returns it as a string. If we were to create a test double that involves this method, the question is what happens if the file doesn't exist at the specified `path`?

- Is it going to return null?

- Is it going to return an empty string?

- Is it going to throw an exception? If so, what is the exception?

The developer writing the test double will do the necessary due diligence to establish the answer to these questions, but they might get it wrong. The previous example was trivial, but you start getting more discrepancies between test doubles and real objects with more complex methods. The way to minimize this problem is to do the proper due diligence when creating the test doubles.

I've heard about London-school TDD and classic-school TDD. What are they?

There is a debate on the internet about which should we use and which is better. London-school TDD focuses on test doubles and it is more suited for business applications. Business applications are apps dealing with DB and UI. The classic-school TDD is more suitable for an algorithmic type of coding.

In this book, we only address the London-school TDD, as we are working on developing business applications.

Why do some developers dislike unit testing and TDD?

Unit testing adds time and complexity to product development, obviously for a good reason, nevertheless, it is still overhead. Unit testing has four major drawbacks:

1. **Development time**: Adding unit testing will multiply development time by multiple factors. Developers under pressure to deliver features as soon as possible will find unit testing overwhelming.

2. **Modifying existing features**: It will require updating the unit tests as well. This can be minimized, but it is hard to eliminate without a major shift from unit testing to Sintegration testing, which is going to be discussed in the next section.

3. **Usage of test doubles**: Some developers have strong feelings against using test doubles, as they tend to produce less realistic tests if the test doubles were not coded correctly.

4. **Unit testing is challenging**: It demands advanced coding skills and coordination between team members, which requires synergy.

Unit testing is not for the fainthearted. I sympathize with these views, but at the same time, I know that companies who are looking for a high-quality product should allocate more time for unit tests. *Drawbacks 2* and *3* can be addressed by better coding practices, which also require more effort.

What is the relationship between TDD and Agile XP?

There are many agile workflow flavors, the most popular ones are **Agile Scrum** and **Agile Kanban**. However, there is one that is less popular and it is software engineering-focused. It is called **Agile XP**, where **XP** stands for **Extreme Programming**.

XP puts unit testing in general, and TDD in particular, at the forefront of its practices, while the other popular agile practices do not go to this level of technical details. XP also tries to solve common software engineering problems, such as project management, documentation, and knowledge sharing.

Code documentation via unit tests

XP believes that the best way to document the code is by having unit tests attached to it versus having some document written elsewhere that would become quickly out of sync with the code. On the other hand, unit tests reflect the current state of the system, as they are regularly checked and updated.

The developer can understand the details of any business rule by looking into the unit tests rather than reading the documentation that may not go to that level of detail.

Can a system survive without TDD?

I would bounce the question back to the developer and ask, do you require the added quality of TDD? Do you agree with the fact that, as the software grows and as the team changes, you need a quality guard?

Yes, you can survive without TDD and unit tests, but the quality may suffer.

There are successful software systems rolled in corporates without unit tests. This is a fact. However, the teams behind them have a higher maintenance overhead and possibly slower release pace, and maybe they have dedicated staff for bug fixing, and possibly follow a waterfall SDLC. This is fine as long as the organization is happy with the quality and cost and they consider the system as *successful*.

TDD with Sintegration tests

One criticism of unit testing is that the unit test code will be tightly coupled with the implementation. Changing production code will have a rippling effect that will force updating, adding, and removing multiple unit tests.

These are ways to reduce coupling with unit tests, which are discussed in *Chapter 6, The FIRSTHAND Guidelines of TDD*, in *The single-behavior guideline* section. However, the provided solutions do reduce coupling but don't get rid of them completely.

On the other hand, integration tests have a dependency on the input and output of the tested functionality. If we are doing integration testing for an API, then we are concerned about what parameters we pass to the API and what we get back, that is, the input and output. This creates loose coupling with the code. Here is a reminder of how integration testing and unit testing operate:

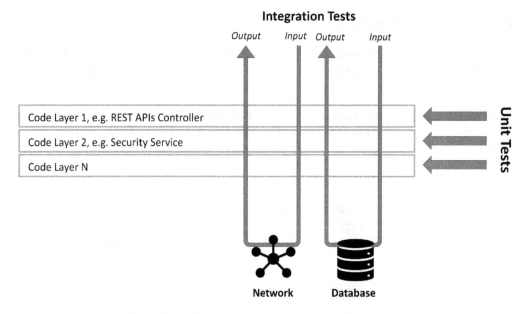

Figure 5.9 – Integration tests versus unit tests

As you can see, unit tests have to understand some details of the layers, while integration tests are interested in the input and output. This is why unit tests have more coupling to the implementation details.

Integration tests come with their own disadvantages, but Sintegration tests solve some of the drawbacks, as we discussed in *Chapter 4, Real Unit Testing with Test Doubles*.

Sintegration testing as an alternative to unit testing with TDD

In recent years, Sintegration tests started competing with unit tests as a way to address these two issues of unit testing:

- Testing from the end-user perspective (the user might be a software client, not necessarily a human). This is also known as **outside-in testing**.

- Having a loose coupling with the code.

Applying TDD principles to Sintegration can work fine in the following ways:

- Test first can be applied in the same way it is applied with unit testing.

- The red, green, refactor approach can work similarly to unit testing.

- Design from the test perspective works in the same way as in unit testing.

- Sintegration can use the same mocking framework to build fakes.

- Sintegration can use the same AAA and method naming conventions.

The main disadvantage is that, while unit testing gives fast feedback as it is focused on small SUTs, the Sintegration tests would not give as fast feedback to the developer. The reason is that they involve building multiple components for the whole feature before a Sintegration test can pass.

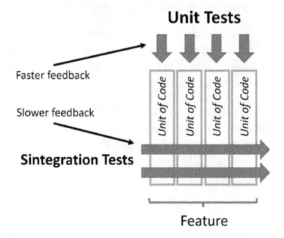

Figure 5.10 – Faster and slower feedback

In the previous figure, you can see the unit tests are only operating on small pieces of code units, and while building these units, you can get the results of your unit test immediately. On the other hand, the feedback of the Sintegration test will be when the whole feature is implemented.

Consider this example. Let's say that you are writing a feature to update a username. The feature would include, but not be limited to, the following units of code:

- Checks username length
- Checks username illegal characters
- Checks whether the user has the right to change the username
- Checks whether the username is already in use
- Gives the user an alternative view if they supply a used username
- Saves username to the database
- Confirms to the user that their username has changed

While, in theory, each of these units of code can have multiple unit tests, and by the time each is coded you get the feedback, the Sintegration test needs to wait until the end of the feature so you can get feedback.

Challenges of Sintegration testing

Sintegration tests still rely on fakes, which are test doubles. Fakes are harder to build and maintain than doing mocks or stubs. The way to master building fakes requires experience in building mocks and stubs, as fakes are usually more complex and require advanced coding.

Also, creating fakes has a time overhead and will delay the start of the project as, before you can write your first Sintegration test, all your related fakes should be ready. For example, if your Sintegration test involves accessing a document database and cloud storage, you might need to create fakes for these components first, before you do any useful Sintegration test.

Practicing TDD with Sintegration requires more experience than practicing it with unit tests. However, the good news is that following the content of this book will help you progress, so when the time comes, and you, with your team, decide to focus on Sintegration, you will have gained the necessary experience to do so.

Summary

We've been through the basics and the principles around TDD, thus, I trust that at this point, you can confidently describe the process to a colleague. However, this chapter is the start of learning TDD, as the book will continue adding to your knowledge as you go.

I held my pen (okay, my keyboard) back from writing more advanced examples and stopped here for a smoother introduction. I hope that I explained the concept in a clear way and encouraged you to continue the book, as the upcoming chapters will have more pragmatic examples that will give you the experience to apply TDD in your own project.

In the next chapter, we will discuss the TDD guidelines and what I refer to as the FIRSTHAND guidelines. You will understand why testing first is important and what value it offers you.

Further reading

To learn more about the topics discussed in the chapter, you can refer to the following links:

- *Martin Fowler on Test-Driven Development*: `https://martinfowler.com/bliki/TestDrivenDevelopment.html`

- *Extreme Programming*: `https://www.agilealliance.org/glossary/xp`

- *Classic TDD or "London School"?*: `http://codemanship.co.uk/parlezuml/blog/?postid=987`

6

The FIRSTHAND
Guidelines of TDD

TDD is more than a test-first unit testing or a Red-Green-Refactor approach. TDD includes best practices and guidelines that steer the way you work with unit testing.

I wanted to make a memorable list, from my experience, of the most useful guidelines on unit testing and TDD. So, here are nine proven best practices that I've abbreviated as **FIRSTHAND**. FIRSTHAND stands for:

- *First*

- *Intention*

- *Readability*

- *Single-Behavior*

- *Thoroughness*

- *High-Performance*

- *Automation*

- *No Interdependency*

- *Deterministic*

In this chapter, we will go through each of these nine guidelines and support them with relevant practical examples. By the end of the chapter, you should have a fair understanding of the ecosystem of TDD and its guidelines.

Technical requirements

The code for this chapter can be found in the following GitHub repository:

`https://github.com/PacktPublishing/Pragmatic-Test-Driven-Development-in-C-Sharp-and-.NET/tree/main/ch06`

The First guideline

Unit tests should be written first. This might seem odd or unintuitive at the beginning, but there are valid reasons for this choice.

Later means never

How many times have you heard, *we'll test it later*? I have never seen a team finishing a project and releasing it to production and then allocating time to unit test their code.

Moreover, adding unit tests at the end will require code refactoring, which might break the product, and it is hard to justify to a non-technical person that a working system was broken because the team was adding unit tests. Actually, the statement *we broke production because we were adding unit tests* sounds ironic. Yes, you can refactor a working system while covered by other types of tests, such as Sintegration and acceptance tests, but it would be difficult to imagine that a team that didn't have time to unit test previously had the time to build other tests that would fully cover the system.

Testing first ensures that unit tests and features are developed hand in hand and tests are not omitted.

Being dependency injection ready

When you get used to the modern software development style of creating a service and then injecting it, you will never look back. Software frameworks have evolved to make DI a first-class citizen. Here are a few examples:

- **Angular Web Framework**: You can only obtain services in Angular via DI and you would struggle to do it in other ways.

- **Microsoft MAUI**: MAUI is a revamp of Xamarin.Forms and one of the major changes from Xamarin is having DI as a first-class citizen.

- **.NET Core Console**: classical .NET Framework console applications did not support DI, but this is now natively supported in Core, which paved the way to have other libraries, built on top of console applications, to support DI, for example ASP.NET Core.

- **ASP.NET Core**: One major difference between classical ASP.NET and ASP.NET Core is having DI as a first-class citizen.

These are all strong signals to tell you that there is no escape from using DI. Having your software implemented and then adding DI later is going to require a major refactoring and rethinking everything.

Starting with unit testing will enforce DI from the first moment.

Designing from the client's perspective

TDD encourages you to think from the client's (the caller) needs rather than getting bogged down with the implementation details. You are encouraged to think of the OOP design such as class name, abstractions, deciding the method signature and return type before thinking of the implementation details like the method body.

If you have a public interface (a combination of classes and methods) used by other systems or libraries, it is harder to change that interface as it is probably in-use by the other system, compared to changing the implementation of your code, say for optimization purposes.

TDD enforces designing the code from the client's perspective.

Promoting behavior testing

A unit test should care about *what* a **System Under Test (SUT)** does, *not how* it does it. In a unit test, you want to push a certain input, check how the dependencies are affected, and check the output. What you should not check is how the SUT worked internally to do all that.

If you decide to check the internals of the SUT, your unit tests will become tightly coupled to the implementation details. This means every change in the method will have a rippling effect on the associated unit tests. This will lead to more unit tests and brittle tests. It is worth iterating here that *tests are an asset and a liability*. Having more tests, which are often unnecessary, means more maintenance.

Testing first makes you naturally think about the inputs, outputs, and side effects, not about the details of the SUT implementation. Testing after implementation leads to what I refer to as *cheating*, where the developers look at the SUT implementation code and write their tests accordingly. This might inadvertently lead to testing implementation details.

TDD promotes the unit testing mantra: Test behavior, not implementation details.

Eliminating false positives

A false positive is when a test passes for the wrong reason. This doesn't happen frequently, but when it does happen, it is hard to catch.

TDD uses the red-green approach to eliminate false positives.

Eradication of speculative code

We have all written code thinking, *perhaps we will need it in the future,* or *let me leave it here as other colleagues will find it useful.* The drawback of this approach is this code may never be used but will be maintained. Even worse, if it is used in the future, it may give the illusion that it has been tested, while, in fact, it has been waiting for a future developer to test it.

TDD eradicates speculative code by writing production-only code.

The Intention guideline

When your system grows, it drives more unit tests that will naturally cover system behavior and documentation. And with more tests comes greater responsibility: **readability** and **maintenance**.

The tests will grow in quantity to an extent where the team will not remember the reason for writing them. You will be looking at a failing test and scratching your head for clues about the intention of the test.

Your unit tests should be understood with the least possible time and effort; otherwise, they will be more of a liability than an asset. An agile software team should be prepared in advance for such test failure scenarios. Intention can be demonstrated by having a clear method signature and a well-structured method body.

Starting with the method signature, here are two popular conventions that should clarify the unit test's intention.

Method_Condition_Expectation

I have been using this convention in naming the unit test methods across the book: `Method_Condition_Expectation`. This is a succinct naming convention that doesn't allow innovative method names and, in my opinion, spares innovation for other tasks. It results in a more boring but standard method name. The following is an example:

`LoginUser_UsernameDoesntExist_ThrowsInvalidOperationException`

This is still not a precise convention, but it is good enough. For example, some developers may argue against using the word `Throws` as it is obvious because the word `Exception` is used.

What is important here is establishing, in a short amount of time, from the three parts seen in the preceding method name, the intention of this test.

> **Note**
> I have seen teams omitting the `Method` part and just using `Condition_Expectation`, especially if the whole unit test class targets one method only.

Method_Should_When

Another popular convention that allows more natural language is using the Method_Should_When. This convention is more akin to fluent coding, where the code flows like an English sentence. The following is an example:

```
LoginUser_Should_Throw_InvalidOperationException_When_
UsernameDoesntExist
```

Sometimes, advocates of this convention like to also use **fluent assertions** for asserting:

```
// Assert
IsSaved.Should().BeTrue();
```

As you might have noticed, the previous code is different from the xUnit style used in the book:

```
// Assert
Assert(true, IsSaved);
```

If you are interested in using fluent assertions, you can have a look at a .NET library called **Shouldly** (`https://github.com/shouldly/shouldly`). In *Appendix 1, Commonly Used Libraries with Unit Tests*, we discuss a similar library called *Fluent Assertions*.

The next part for clarifying the intention is clarifying the method body.

Unit test structure

The dominant method for structuring a unit test body is the popular **Arrange-Act-Assert (AAA)**. Let's shed more light on the intention of what each section should do.

Arrange

Arrange is meant to achieve two objectives:

- Initializing variables
- Creating a SUT state

The Arrange section might be shared with the constructor of the unit test class as the constructor might be doing some preparations to reduce the Arrange code across each unit test. In other words, there might be some *arrangements* happening outside this section. We will see examples of this in *Part 2, Building an Application with TDD*, of the book.

Initializing the objects and initializing expectations happen in the Arrange section, and that is the obvious part. The not-so-obvious part is that this section sets a *state*. Here is an example to clarify what a state means:

```
LoginUser_UsernameDoesntExist_ThrowsInvalidOperationException
```

Arrange will create a state where the user doesn't exist in the system, and in most cases, that code will be tightly related to the Condition in Method_Condition_Expectation. In this case, the arrangement should be linked to UsernameDoesntExist.

Act

Act is mostly a single line of code that calls the same method specified in the first part of the method name. In the previous example, I expect the act to look like this:

```
// Act
var exception = Record.Exception(() => sut.LoginUser(…));
```

This method reiterates the intention specified in the method's signature and provides clean code.

Assert

Assert is asserting the conditions in the last section of the method signature: Expectation.

The signature convention and the body structure work together to provide clear intentions. Keep in mind that no matter what convention you've used, the key is consistency and clarity.

Clear-intention unit tests would promote easier maintenance and more accurate documentation.

The Readability guideline

Is this method readable? Do you need to run it and start debugging to understand what it does? Does the Arrange section make your eyes bleed? This might be violating the readability principle.

Having the Intention guideline established is fabulous, but it is not enough. You will have at least 10x more lines of code in your unit tests compared to your production code. All this needs to be maintained and grow with the rest of your system.

Tidying up the unit test for readability follows the same practices as the production code. However, there are some scenarios that are more dominant in unit tests, which we are going to address here.

SUT constructor initialization

Initializing your SUT will require that you prepare all the dependencies and pass them to the SUT, something like this:

```
// Arrange
const double NEXT_T = 3.3;
const double DAY5_T = 7.7;
var today = new DateTime(2022, 1, 1);
```

```
var realWeatherTemps = new[]
    {2, NEXT_T, 4, 5.5, 6, DAY5_T, 8};
var loggerMock =
    Substitute.For<ILogger<WeatherController>>();
var nowWrapperMock = Substitute.For<INowWrapper>();
var randomWrapperMock = Substitute.For<IRandomWrapper>();
var clientMock = Substitute.For<IClient>();
clientMock.OneCallAsync(Arg.Any<decimal>(), Arg.Any<decimal>(),
    Arg.Any<IEnumerable<Excludes>>(), Arg.Any<Units>())
    .Returns(x =>
    {
        const int DAYS = 7;
        OneCallResponse res = new OneCallResponse();
        res.Daily = new Daily[DAYS];
        for (int i = 0; i < DAYS; i++)
        {
            res.Daily[i] = new Daily();
            res.Daily[i].Dt = today.AddDays(i);
            res.Daily[i].Temp = new Temp();
            res.Daily[i].Temp.Day =
                realWeatherTemps.ElementAt(i);
        }
        return Task.FromResult(res);
    });
var controller = new WeatherController(loggerMock,
    clientMock, nowWrapperMock, randomWrapperMock);
...
```

Now we have done all the coding preparation, we can initialize the SUT (the controller, in our case) and pass to it the right parameters. This will be repeated in most of your same-SUT tests, which will make it a read nightmare. This code can easily go to the constructor of the unit test class and become something like this:

```
private const double NEXT_T = 3.3;
private const double DAY5_T = 7.7;
private readonly DateTime _today = new(2022, 1, 1);
private readonly double[] _realWeatherTemps = new[]
```

```
    { 2, NEXT_T, 4, 5.5, 6, DAY5_T, 8 };

private readonly ILogger<WeatherForecastController> _loggerMock
    = Substitute.For<ILogger<WeatherForecastController>>();
private readonly INowWrapper _nowWrapperMock =
    Substitute.For<INowWrapper>();
private readonly IRandomWrapper _randomWrapperMock =
    Substitute.For<IRandomWrapper>();
private readonly IClient _clientMock =
    Substitute.For<IClient>();
private readonly WeatherForecastController _sut;

public WeatherTests()
{
    _sut = new WeatherForecastController(_loggerMock,
        _clientMock, _nowWrapperMock, _randomWrapperMock);
}
```

The beauty of the previous code is that it is reusable by all unit tests in the same class. The code in your unit test becomes something like this:

```
// Arrange
_clientMock.OneCallAsync(Arg.Any<decimal>(),
    Arg.Any<decimal>(),
    Arg.Any<IEnumerable<Excludes>>(), Arg.Any<Units>())
    .Returns(x =>
    {
        const int DAYS = 7;
        OneCallResponse res = new OneCallResponse();
        res.Daily = new Daily[DAYS];
        for (int i = 0; i < DAYS; i++)
        {
            res.Daily[i] = new Daily();
            res.Daily[i].Dt = _today.AddDays(i);
            res.Daily[i].Temp = new Temp();
            res.Daily[i].Temp.Day =
                _realWeatherTemps.ElementAt(i);
```

```
        }
        return Task.FromResult(res);
    });
...
```

The `Arrange` size in your unit tests in this class has gone down. Remember, you are looking here at one unit test method, but you might have several unit tests for the same SUT.

You might argue that while we have cleared some of the repeated code in our `Arrange`, it is still busy. Let's get the builder design pattern to the rescue.

The builder pattern

For the same SUT, every unit test's arrangement varies slightly from others. The **builder design pattern** is useful when creating an object with lots of possible configuration options, which comes in handy for this scenario.

> **Note**
> This is different from the **Gang of Four (GoF)** Builder design pattern.

The builder class, for the previous example, looks like this:

```
public class OneCallResponseBuilder
{
    private int _days = 7;
    private DateTime _today = new (2022, 1, 1);
    private double[] _temps = {2, 3.3, 4, 5.5, 6, 7.7, 8};

    public OneCallResponseBuilder SetDays(int days)
    {
        _days = days;
        return this;
    }

    public OneCallResponseBuilder SetToday(DateTime today)
    {
        _today = today;
        return this;
```

```
        }

        public OneCallResponseBuilder SetTemps(double[] temps)
        {
            _temps = temps;
            return this;
        }

        public OneCallResponse Build()
        {
            var res = new OneCallResponse();
            res.Daily = new Daily[_days];
            for (int i = 0; i < _days; i++)
            {
                res.Daily[i] = new Daily();
                res.Daily[i].Dt = _today.AddDays(i);
                res.Daily[i].Temp = new Temp();
                res.Daily[i].Temp.Day = _temps.ElementAt(i);
            }
            return res;
        }
}
```

What is notable in this class is that:

1. Every method returns the class instance. This helps to chain methods like this:

```
OneCallResponse res = new OneCallResponseBuilder()
    .SetDays(7)
    .SetTemps(new []{ 0, 3.3, 0, 0, 0, 0, 0 })
    .Build();
```

2. The Build() method will combine all the configurations together to return a usable object.

The refactored Arrange of the previous unit test looks like this:

```
// Arrange
OneCallResponse res = new OneCallResponseBuilder()
    .SetTemps(new []{ 0, 3.3, 0, 0, 0, 0, 0 })
```

```
    .Build();

_clientMock.OneCallAsync(Arg.Any<decimal>(),
    Arg.Any<decimal>(), Arg.Any<IEnumerable<Excludes>>(),
    Arg.Any<Units>())
    .Returns(res);
```

The preceding code takes advantage of the builder class that we created earlier. You can see the code clearly setting the next day's temperature as 3.3 degrees.

Using SUT constructor initialization and the builder pattern are just a few examples of making your unit tests readable.

You can find the refactored class in `WeatherForecastTestsReadable.cs` and the original in `WeatherForecastTestsLessReadable.cs` in the companion source code.

Readability promotes the healthy growth of your unit test's codebase. Keep it in check from day one.

The Single-Behavior guideline

Every unit test should test one and only one behavior. Throughout this book, this concept has been enforced naturally by:

- The naming of the unit test method's signature, which reflects one condition with one expectation
- A single AAA structure that enforced a single `Act`

Before digging further, I would like to define the word *behavior*.

What is behavior?

The definition of behavior varies in the industry, so it is important to set an accurate one for the context of this book. Each SUT is supposed to do something. A SUT does this *thing* by:

- **Communicating with dependencies**: Communication can be by calling a method on a dependency or setting a field or a property – this is referred to as *external behavior*.
- **Returning a value to the outside world (the caller)**: This could be via an `Exception` or the return value (if a method is not a `void` or a `Task` method) – this is also referred to as *external behavior*.
- **Plumbing all together**: Doing various commands in preparation to receive the input, in preparation for the output (the return value), or in preparation for communicating with a dependency – this is referred to as the *internals* of the SUT, or *internal behavior*.

External behavior propagates across the system, as it is touching other dependencies, while internal behavior is encapsulated in the SUT and not shown to the outside world.

When we use the word *behavior* on its own, we mean *external behavior*, so the Single-Behavior guideline refers to single external behavior. As usual, definitions seem more complicated than they are, so let's fortify this definition with an example.

Example of behavior

Let's take this code from the **Weather Forecasting App** (**WFA**) that was introduced in *Chapter 2, Understanding Dependency Injection by Example*:

```
public async Task<IEnumerable<WeatherForecast>> GetReal()
{
    const decimal GREENWICH_LAT = 51.4810m;
    const decimal GREENWICH_LON = 0.0052m;
    OneCallResponse res = await _client.OneCallAsync
      (GREENWICH_LAT, GREENWICH_LON, new[]{Excludes.Current,
      Excludes.Minutely, Excludes.Hourly, Excludes.Alerts},
      Units.Metric);

    WeatherForecast[] wfs = new
        WeatherForecast[FORECAST_DAYS];
    for (int i = 0; i < wfs.Length; i++)
    {
        var wf = wfs[i] = new WeatherForecast();
        wf.Date = res.Daily[i + 1].Dt;
        double forecastedTemp = res.Daily[i + 1].Temp.Day;
        wf.TemperatureC = (int)Math.Round(forecastedTemp);
        wf.Summary = MapFeelToTemp(wf.TemperatureC);
    }
    return wfs;
}
private string MapFeelToTemp(int temperatureC)
{
    ...
}
```

In the preceding code, the external behaviors are all in the `_client.OneCallAsync` call and in the `return` statement. The rest of the code is all internal. You can think of the internal code's role as preparing to trigger the `_client` dependency and to return a value.

The internal behaviors are no longer relevant once these two external behaviors are triggered; they are executed and forgotten, while the external behaviors propagate to other services.

Testing external behavior only

If the role of internals is to prepare for external behavior only, then testing external behavior will cover testing the whole code. You can think of it as the internal code being tested is a by-product of testing the external behavior.

Here are some examples of unit testing the behavior (external behavior). These examples are fully implemented in the source code in *Chapter 3, Getting Started with Unit Testing* (**WF** stands for **Weather Forecasting**):

```
GetReal_NotInterestedInTodayWeather_WFStartsFromNextDay


GetReal_5DaysForecastStartingNextDay_
    WF5ThDayIsRealWeather6ThDay


GetReal_ForecastingFor5DaysOnly_WFHas5Days


GetReal_WFDoesntConsiderDecimal_RealWeatherTempRoundedProperly


GetReal_TodayWeatherAnd6DaysForecastReceived_
    RealDateMatchesLastDay


GetReal_TodayWeatherAnd6DaysForecastReceived_
    RealDateMatchesNextDay


GetReal_RequestsToOpenWeather_MetricUnitIsUsed


GetReal_Summary_MatchesTemp(string summary, double temp)
```

After executing these tests that are targeting external behavior only, a code coverage tool will show that all the code that was demonstrated in the previous example is covered by our test. This includes the code in the GetReal() public method and the code in the MapFeelToTemp() private method. Let's have a look at an example of code coverage:

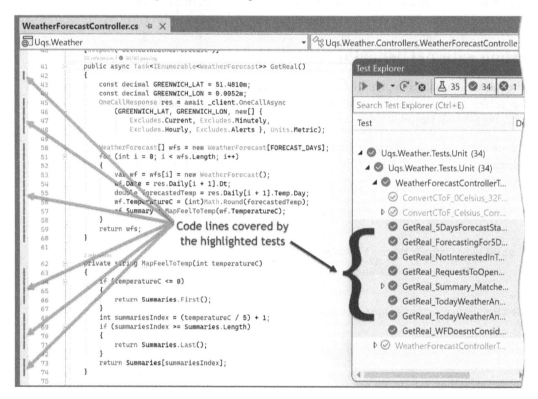

Figure 5.1 – Lines covered by behavior tests

In the preceding figure, I have used a VS plugin called **Fine Code Coverage** (**FCC**) to show the lines that are covered by the selected tests to the right. It shows that all the code lines in the SUT were covered by these tests. We will discuss coverage and this plugin tool in more detail in *The Thoroughness guideline* section.

Why not test internals?

One common mistake in unit testing is when developers try to test the internals of a SUT. Here are a few problems with testing internals:

- The SUT is already covered when testing external behavior, so there's no need to increase the number of unit tests, which will increase maintenance (liability).

- Testing the internals will create tight coupling between the SUT and the tests, which will create brittle tests that will have to change often.

- Internals are usually hidden behind non-public methods; to test them, the code needs to change to public, which violates OOP's encapsulation.

A single behavior per test

Now that the definition of behavior is clear, we can clarify what testing a single behavior means. Testing a single behavior works alongside the Intention guideline. If we are targeting a single behavior, our test will promote a better intention and more readability, and when it fails, we are supposed to pinpoint the reason promptly.

Single-behavior testing is made up of a SUT, a single condition, a single expectation, and minimal assertions. The unit test signatures in the previous list (section: *Testing external behavior only*) are examples of targeting a single behavior. Also, all the examples in this book follow the same guidelines.

A unit test method should test a single behavior and never test internals.

The Thoroughness guideline

When unit testing some naturally occurring questions are as follows:

- How many tests are enough?

- Do we have a test coverage metric?

- Should we test third-party components?

- What system components should we unit test and what should we leave?

The Thoroughness guideline attempts to set the answers to these questions.

Unit tests for dependency testing

When you encounter a dependency, whether this dependency is part of your system or a third-party dependency, you create a test double for it and isolate it in order to test your SUT.

In unit tests, you do not directly call a third-party dependency; otherwise, your code will be an integration test and with that, you lose all the benefits of unit tests. For example, in unit tests, you do not call this:

```
_someZipLibrary.Zip(fileSource, fileDestination);
```

For testing this, you create a test double for the .zip library to avoid calling the real thing.

This is an area that unit tests do not and should not cover, which leaves us with a coverage issue as some areas of the code are not unit-testable.

To test the interaction with dependencies and solve the previous problem of not being able to unit test some code, other types of tests can be employed, such as Sintegration, integration, and acceptance tests.

We started to speak about *coverage*; now we can dig deeper into the topic.

What is code coverage?

The first step in understanding thoroughness is understanding code coverage. **Code coverage** is the percentage of the system code lines executed by your tests (unit, Sintegration, integration, and so on) from the total system lines. Let's assume we have a method that returns if an integer is even:

```
public bool IsEven(int number)
{
    if(number % 2 == 0) return true;
    else return false;
}
```

Let's have a unit test that tests if a number is even:

```
public void IsEven_EvenNumber_ReturnsTrue() {…}
```

This unit test would have covered the `if` line but it wouldn't have executed the `else` line. This constitutes 50% code coverage. Obviously, having another test to test an odd number would have led to 100% coverage.

It is important to realize that code coverage is not necessarily the unit test coverage only, but it might be a combination of unit tests, Sintegration tests, and integration tests. However, out of all tests, unit tests usually cover the biggest chunk of the code, as they are easier to write than other tests. Also, when using the TDD style, you get high coverage as soon as the feature is implemented, as all relevant unit tests are already supplied.

Coverage measurement tools

To measure how much of your code is covered by your tests, usually, you measure it in the **Continuous Integration** (**CI**) pipeline (which will be discussed in *Chapter 11, Implementing Continuous Integration with GitHub Actions*, of the book) and/or on the development machine. There are plenty of commercial options for running test coverage and a limited number of free options. Here are a few examples:

- NCover – commercial
- dotCover – commercial
- NCrunch – commercial
- VS Enterprise code coverage – commercial

- SonarQube – commercial and community

- AltCover – free

- FCC – free plugin for VS

If you want to see how code coverage works, you can install FCC by following these steps:

1. From the menu, select **Extensions | Manage Extensions** and the **Manage Extensions** dialog will open.

2. Select **Online | Visual Studio Marketplace**.

3. Search for `Fine Code Coverage` and then select **Download**.

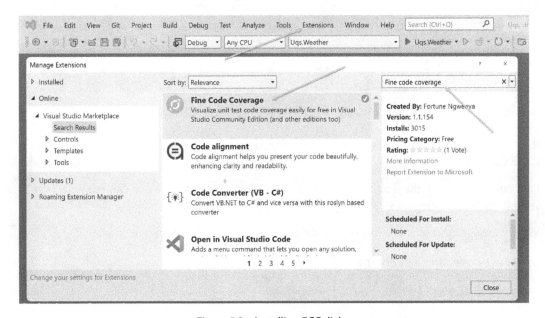

Figure 5.2 – Installing FCC dialog

4. Restart VS.

After installation, you can open the project from *Chapter 3, Getting Started with Unit Testing* and execute all your unit tests (**Test | Run All Tests**). This tool will be triggered automatically after several seconds of executing the tests. To see the results, a panel like the following one will appear at the bottom of VS to show you the results:

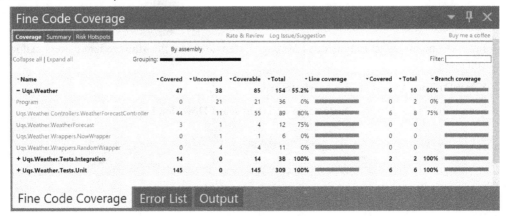

Figure 5.3 – FCC analysis results

If you can't see this panel, you might need to go, from the menu, to **View | Other Windows | Fine Code Coverage**.

I am only interested in the production code cover, which is Uqs.Weather in this example. I can see that my total coverage is 55.2%.

We will speak further about the previous code coverage results in the next few sections.

Unit test coverage spectrum

Unit tests are excellent for testing the following:

- Business logic
- Validation logic
- Algorithms
- Interaction between components (not to be confused with integration between components)

On the other hand, unit tests are not optimal for testing the following:

- Integration between components – the call from component A passing through to component B, such as getting info from the database and analyzing the data
- Service booting components – such as Program.cs
- Direct call to dependencies (dependency testing, as discussed earlier)

- Wrappers wrapping real components – such as RandomWrapper in the following example:

```
public interface IRandomWrapper
{
    int Next(int minValue, int maxValue);
}
public class RandomWrapper : IRandomWrapper
{
    private readonly Random _random = Random.Shared;

    public int Next(int minValue, int maxValue)
    {
        return _random.Next(minValue, maxValue);
    }
}
```

Unit tests should have no interest in testing anything in this class, as this is wrapping the real components directly.

Now, if we go back to the results of the code coverage, it is clear why we have 0% coverage for Program. cs, NowWrapper.cs, and RandomWrapper.cs. It is best that the code in these files is not tested by unit tests and we haven't done that.

WeatherForecastController is 80% covered. You can open the file to see what FCC highlights.

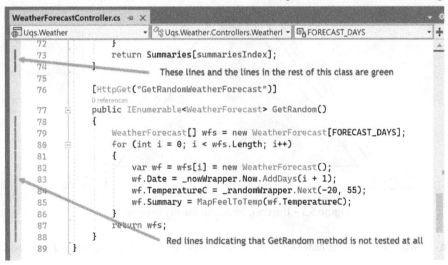

Figure 5.4 – Highlights from FCC

It looks like not even a single line in `GetRandom` is tested as all are *red*. I might have not targeted this method in any of my tests. Obviously, having a complete method that is not tested would not have happened if I were using TDD; however, this is a method that comes with the VS sample code for the ASP.NET project.

Now that we understand what code coverage is and what can be covered, we can describe what being *thorough* in testing is.

Being thorough

Clearly, the best coverage level is 100%; or at least this should be the intention, but it is not easily achievable and sometimes, it isn't worth the cost and effort to achieve it.

First, as discussed, unit tests are not meant to provide 100% coverage, so they need to be backed up by other categories of tests. If we want to achieve 100% coverage using unit tests, we will shoehorn unit tests to test things that are not best suited for unit tests.

Being thorough is doing a combination of unit testing, Sintegration testing, integration testing, and user acceptance testing.

The cost, time, and quality triangle

This is probably not the first time you've heard of this concept. It is a project management one and is not specific to software engineering. Here is the popular triangle:

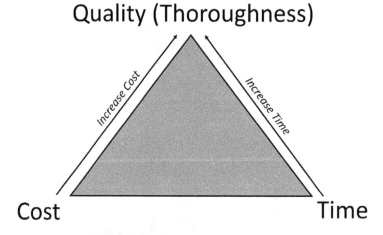

Figure 5.5 – The cost, time, and quality triangle

Quality in our triangle represents how thorough we want to be. It is clearly a function of **Time** and **Cost**.

Testing, in general, is a time-consuming process and unit testing takes as much time, if not more, as writing production code. Unit tests are written by the same developers as the production code and they are not usually things that are done in parallel; they are done in series. The red-green-refactor process of TDD is done by the same individual or the team writing the code, not by a separate tester doing this in parallel.

How thorough?

Aiming for over 80% coverage is good. 95% coverage is the max you would be able to have with reasonable effort. This is only the case for the combination of unit testing with Sintegration or unit testing with integration. Acceptance testing does not usually count toward your coverage.

Being more thorough and aiming for higher coverage is a matter of more time and cost. So, the question of how much coverage is a question for project management and your team.

Being thorough is aiming for high coverage using a combination of unit tests with Sintegration tests or unit tests with integration tests. This is taking into consideration the time, money, and quality triangle.

The High-Performance guideline

Your unit tests should not take, in today's hardware, over 5 seconds to run, ideally no more than a couple of seconds after the tests are loaded. But why all this fuss? Can't we just let them take whatever time is needed to run without sweating over it?

First, your unit tests will have to run many times throughout the day. TDD is about running a chunk of your unit tests or all of them with every change; therefore, you don't want to spend your time waiting and lose valuable time that could be spent more productively.

Second, your unit tests need to provide fast feedback to your CI pipeline. You want your source control branches to be green all the time, so that other developers are pulling green code at any given time and, of course, it is ready to ship to production. This is even more important for larger teams.

So, how do you keep your unit tests performing as fast as possible? We will attempt to answer this question in the next sections.

Integration as disguised units

I have seen this in many projects, where developers call their integration tests unit tests just because they are executed by NUnit or xUnit.

Integration tests are slow by nature as they are doing IO operations, such as writing or reading from the disk, going to the database, and going over the network. These operations consume time and result in tests that take minutes or more to run.

Unit tests use test doubles and rely on memory and CPU to run. Running 10K unit tests should take seconds after project load, so *make sure you are not executing integration tests.*

CPU- and memory-intensive unit tests

You might be doing unit tests that rely on math libraries that are not treated as test doubles or you might have code that involves sophisticated logic.

You can have multiple categories of unit tests that can be executed at different times. xUnit calls this feature a **trait**. You can have a trait to indicate slow tests.

You can execute faster tests during your TDD and all tests before pushing to the codebase (source control).

Having too many tests

Let's assume that after becoming a TDD guru and reaching 100K tests, you start having tests that are taking over 5 seconds.

I would directly ask:

- *Do you have a bloated monolith type of project?*
- *What is your problem: slow tests or a project that should be divided into microservices?*

A temporary solution for this issue is having multiple VS solutions and figuring out how to divide related projects into various solutions.

Having too many tests is a sign of weak architecture. Employing an alternative architecture, such as microservices, can be considered and the unit tests problem may be sorted out automatically.

High-performance unit tests are something you strive for from day one and keep in check while building your project.

The Automation guideline

When we say automation, we mean CI. CI has a dedicated chapter in this book, *Chapter 11, Implementing Continuous Integration with GitHub Actions,* so we won't go into the details here.

This guideline is about realizing that the unit tests will run on other platforms than your local development machine. So, how do you make sure your unit tests are ready for automation?

CI automation from day 1

Agile teams dedicate the first sprint to setting up the environment, including the CI pipeline. This is usually called sprint or iteration zero. If CI is set up from day 1 to listen to source control, there is less chance it is omitted or a CI-incompatible test is introduced.

Implement CI from the start of the project.

Platform-independent

.NET is multi-platform and the trend nowadays is to use Linux servers to run CI pipelines. Also, the developer dev machine can be Windows, macOS, or Linux.

Make sure your code does not rely on any OS-specific features, if not really needed, so that you can have a free choice of which OS to use for the CI pipeline.

High performance on CI

CI pipelines, nowadays, are leased services from CI providers such as Azure DevOps or GitHub Actions. These leased services are low in resources (CPU and memory) and are shared between multiple projects.

It is fair to say that tests will take double the time, if not more, to run on a CI pipeline in comparison to your local machine, assuming that you have a decent dev machine and you are not paying a fortune for your CI resources.

This reiterates the fact that having high-performing tests is necessary.

Make sure whatever test you write on your dev machine is ready to run on CI.

The No Interdependency guideline

First, I would like to elevate this from a guideline to a principle. This principle ensures that the unit test does not alter a state permanently; or, in other words, executing a unit test should not persist data. Based on this principle, we have the following rules:

- Test A shouldn't affect test B.
- It doesn't matter whether test A runs before test B.
- It doesn't matter whether we run test A and test B in parallel.

If you think about it, a unit test is creating test doubles and doing its operations in memory, and as soon as it finishes execution, all the changes are lost, except the test report. Nothing is saved in the database, in a file, or anywhere, because these dependencies were all provided as test doubles.

Having this principle in place also ensures that the test runner, such as Test Explorer, can run the tests in parallel and use multi-threading if needed.

Ensuring this principle is a shared responsibility between the unit testing framework and the developer.

Unit testing framework responsibility

The unit testing framework should make sure that the unit test class is re-initialized after executing each unit test method. A unit test class should not maintain the state in the same way that a normal class maintains the state. Let me demonstrate with an example:

```
public class SampleTests
{
    private int _instanceField = 0;
    [Fact]
    public void UnitTest1()
    {
        _instanceField += 1;
        Assert.Equal(1, _instanceField);
    }
    [Fact]
    public void UnitTest2()
    {
        _instanceField += 5;
        Assert.Equal(5, _instanceField);
    }
}
```

When a unit testing framework runs UnitTest1 and UnitTest2, it should not run them from the same object; it creates a new object for each method run. This means that they do not share instance fields. Otherwise, if UnitTest1 runs before UnitTest2, then we should be asserting for 6.

> **Important Note**
>
> Modifying an instance field in a method, then asserting it in the same method, is not a good unit testing practice, but it has been done for demonstration purposes.

xUnit does follow this principle of making sure the state is not shared. However, it can be instructed otherwise by using a static member.

Developer's responsibility

As a developer, you know that you should not persist data during a unit test. Otherwise, this is, by definition, not a unit test.

If a developer is using test doubles to replace the database, the network, and other dependencies, they are respecting this rule by virtue. However, the problem arises if the developer decides to use static fields. Static fields will preserve the state between independent method calls. Let's take this example:

```
public class SampleTests
{
    private static int _staticField = 0;
    [Fact]
    public void UnitTest1()
    {
        _staticField += 1;
        Assert.Equal(1, _staticField);
    }
    [Fact]
    public void UnitTest2()
    {
        _staticField += 5;
        Assert.Equal(6, _staticField);
    }
}
```

The previous code fails this rule, as running UnitTest2 before UnitTest1 will fail the test. Actually, there is no guarantee that the second method will run after the first one based on their order in the class.

There are instances where using static fields would promote better performance. Say we have an in-memory read-only database that we need to use in all tests. Also assume that this is taking a long time to initialize, such as 100 ms:

```
private readonly static InMemoryTerritoriesDB =
    GetTerritories();
```

The field is read-only and the content of the class is also read-only, such as readonly record. Although, for this scenario, I argue if you need all the territories, or can you create a cutdown version suitable for unit testing? This might speed up loading and can remove the need for a static field.

As a developer, you should be careful not to create a state between multiple unit tests. This is easy to miss if you are new to the unit testing field.

No interdependency leads to easier-to-maintain code and reduces bugs in unit tests.

The Deterministic guideline

A unit test should have a deterministic behavior and should lead to the same result. This should be the case regardless of the following:

- **Time**: This includes changes in time zone and testing at different times.

- **Environment**: Such as the local machine or CI/CD server.

Let's discuss some cases where we risk making non-deterministic unit tests.

Non-deterministic cases

There are cases that can lead to non-deterministic unit tests. Here are a few of them:

- Having interdependent unit tests, such as a test that writes to a static field.

- Loading a file with an absolute path as the file location on the development machine will not match that on the automation machine.

- Accessing a resource that requires higher privileges. This can work, for example, when running VS as an admin but may fail when running from a CI pipeline.

- Using randomization methods without treating them as dependencies.

- Depending on the system's time without treating it as a dependency.

Next, we will see a case of not depending on varying time, to make our unit test deterministic.

Example of freezing time

If your test relies on time, you should be using a test double to freeze the time to ensure a deterministic test. Here is an example:

```
public interface INowWrapper
{
    DateTime Now { get; }
}
public class NowWrapper : INowWrapper
{
    public DateTime Now => DateTime.Now;
}
```

This is a wrapper to allow injecting the current time as a dependency. To register the wrapper in `Program.cs`:

```
builder.Services.AddSingleton<INowWrapper>(_ => new
    NowWrapper());
```

Your service can look like this:

```
private readonly INowWrapper _nowWrapper;
public MyService(INowWrapper nowWrapper)
{
    _nowWrapper = nowWrapper;
}
public DateTime GetTomorrow() =>
    _nowWrapper.Now.AddDays(1).Date;
```

The calculation in the preceding code is for demo purposes and does not take into account daylight saving. To inject the current time from the unit tests:

```
public void GetTomorrow_NormalDay_TomorrowIsRight()
{
    // Arrange
    var today = new DateTime(2022, 1, 1);
    var expected = new DateTime(2022, 1, 2);
    var nowWrapper = Substitute.For<INowWrapper>();
    nowWrapper.Now.Returns(today);
    var myService = new MyService(nowWrapper);

    // Act
    var actual = myService.GetTomorrow();

    // Assert
    Assert.Equal(expected, actual);
}
```

The previous unit test has frozen the current time to a specified value. This made the code independent from the OS's clock and thus made it deterministic.

Running a unit test should always yield the same results, regardless of time or environmental factors.

Summary

FIRSTHAND accumulates valuable guidelines and best practices in the industry. I trust this chapter topped up the learnings of the previous chapters to help you understand TDD and its ecosystem. I also hope that it made these guidelines memorable as TDD comes up often in developer discussions and it is certainly likely to be an interview topic.

This chapter marks the end of this section, where we looked at dependency injection, unit testing, and TDD. This section was only an introduction to TDD, with scattered small and mid-size examples. If you've made it to this point, then hats off, you have covered the basics of TDD.

The next section will take all the basics and apply them to more lifelike scenarios. To make sure that you are ready for this application and to mimic a realistic application that uses TDD, our next chapter will be about **domain-driven design** (**DDD**) as you will be using the DDD concepts in later chapters.

Part 2:
Building an Application
with TDD

TDD is usually combined with a **Domain-Driven Design (DDD)** architecture. In this part, we will take what we've learned in *Part 1* and use it to build a complete application using TDD and DDD; in one instance, we will use a relational DB (*Chapter 9*), and in another, a document DB (*Chapter 10*) to show how this would affect our unit test implementation.

By the end of this part, you should be able to build an application using TDD and DDD from scratch. The following chapters are included in this part:

- *Chapter 7, A Pragmatic View of Domain-Driven Desig*
- *Chapter 8, Designing an Appointment Booking App*
- *Chapter 9, Building an Appointment Booking App with Entity Framework and Relational DB*
- *Chapter 10, Building an App with Repositories and Document DB*

7
A Pragmatic View of Domain-Driven Design

Domain-driven design (DDD) is a set of software design principles that are widely used in modern enterprise applications. They were bundled and made popular in 2003 by Eric Evans in his book *Domain-Driven Design*.

You might be wondering how this is related to **test-driven development** (TDD). Is it because it is a similar-sounding acronym? The reality is that TDD and DDD work together where TDD covers the design from the client's perspective and the quality, while DDD complements the rest of the design. You will hear the two terms used together in a conversation and in job specifications, and the reason for this will be clear by the end of *Part 2, Building an Application with TDD*.

This chapter is meant to be a primer on DDD, so you will have the foundation required to build a complete application using a combination of TDD and DDD.

DDD is a technical and a philosophical topic. Given the pragmatism of this book and the length of this chapter, our focus will be limited to the pragmatic aspect of DDD related to the application we are implementing in the following chapters.

In this chapter, we will cover the following topics:

- Working with a sample application
- Exploring domains
- Exploring services
- Exploring repositories
- Putting everything together

By the end of the chapter, you will understand the basic DDD terminology and be able to explain it to a colleague.

This is usually transferred in JSON. This is an example of the preceding C# contract being serialized as JSON:

```json
{
    "id": 1,
    "content": "Some content",
    "author": {
        "id": 100,
        "name": "John Smith"
    },
    "createdDate": "2022-01-01T01:01:01",
    "numberOfComments": 5,
    "numberOfViews": 486,
    ...
}
```

Contracts are not part of the DDD philosophy, but they are needed here to have a complete application.

The domain layer project

The components of this layer are in `Uqs.Blog.Domain`. This is where all the types related to the domain design live.

> **Important Note**
>
> Dividing layers, naming projects, and arranging them based on layers is a highly opinionated process. There is no widespread industry standard of the best approach. So, consider my approach here as *an example* rather than *the way* to do it.

This layer contains the following:

- Business logic
- Database persistence

Our project resembles a similar design to this one:

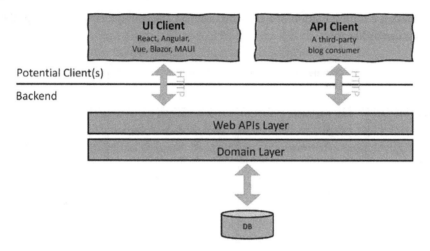

Figure 7.3 – A design diagram for the application

This diagram represents our application; however, it's worth noting that DDD is concerned with the backend, not the client.

Next, we will go through the constituents of DDD and we will start with the domain.

Exploring domains

DDD is a collection of software design philosophies and best practices. There are a handful of books dedicated to DDD, and most of them are above 500 pages. So, we can talk a lot about DDD, but this book isn't about DDD, so we will be brief.

DDD focuses on business logic and the interaction with the DB and the outside world and employs a set of practices for a robust software design. The word *domain* in DDD refers to a *business domain*, which can be car insurance, accounting, billing, banking, e-commerce, and others. DDD emphasizes the business domain, as per the term *domain-driven*.

Next, we will explore the architectural components that make the practical aspect of DDD.

Entities

An object primarily defined by its identity is called an **entity**. It is a type of domain model that needs to be tracked over time and whose attributes are likely to change over time. A perfect example of this would be a person entity, which has a changeable email and home address but a fixed identity, which is the person herself/himself.

In our preceding blog example, `Post`, `Author`, `Comment`, and `Commenter` are entities.

`Comment` is peculiar as some can argue that it is a value type! But what if it is editable? Then, its identity becomes important.

Entities are represented as classes and records, and they definitely have an **identifier** (ID).

Entity versus value objects

When designing your domain, it is important to understand the differences so that you pick the right design. Here are the major distinguishing aspects:

- **Lifespan**: Entities live in a continuum while value objects are created and destroyed with ease.

- **Immutability**: An object is said to be immutable if its value cannot change after creation. Entities are mutable while value objects are immutable.

- **Identifier**: Entity objects require an identifier while value objects don't.

- **Classes or structs**: Entities use classes and adhere to the .NET reference type principles (stored in heap, passed by reference, and so on) while value types are structs (at least as DDD recommends), which adhere to the .NET value type principles (stored in stack, passed by value, and so on).

To summarize, when we design our domain objects, they can be done as entities or value objects depending on whether they represent an identity.

Aggregates

Aggregates are a group of classes that form one business aim. The previous blog classes set a distinguished business objective, which is managing a blog post. These classes form an aggregate.

> **Important Note**
> The *aggregation* term that is used in **object-oriented programming** (**OOP**) and the **Unified Modeling Language** (**UML**) is not the same concept as the DDD aggregate.

An **aggregate root** is the single main entity of an aggregate. DDD recommends that accessing (invoking a method) any domain object in an aggregate is conducted through the aggregate root. It is clear in the blog example that the aggregate root is the `Post` domain object.

Anemic models

When we studied OOP, we learned that an object handles its own data and its behavior. So, if we have a class called `Person`, there might be a read-only property called `Email` in that class. Also, to set the email address, you will have a method that might be creatively called `void ChangeEmail(string email)`, which does some business logic and validations before setting the email. Our class, according to DDD, would look like this:

```
public class Person
{
    public string Email { get; private set; }
    public void ChangeEmail(string email)
    {
        ...
    }
    // other properties and methods
}
```

This class stores its own data. For example, the `Email` property is storing the email value and there is a behavior, which is represented by the `ChangeEmail` method, which is changing the stored `Email`.

An **anemic model** is an object that contains little or no behavior and focuses on carrying data. Let's give an example of an anemic model by transferring our `Person` class earlier to an anemic version:

```
public class Person
{
    public string Email { get; set; }
    // other properties
}
```

Now the email has a setter, but how are validation and other business logic implemented if it is not within the class itself? The answer is that another class would be responsible, such as the following:

```
public class PersonService
{
    public void ChangeEmail(int personId, string email)
    {
        Person person = ...; // get the object some how
        // validate email format
        // check that no other person is using the email
        person.Email = email;
```

```
    {
        var author = _authorRepository.GetById(authorId);
        if (author is null)
        {
            throw new ArgumentException(
              "Author Id not found",nameof(authorId));
        }
        if (author.IsLocked)
        {
            throw new InvalidOperationException(
              "Author is locked");
        }
        var newPostId = _postRepository.CreatePost
          (authorId);
        return newPostId;
    }
}
```

First, you'll notice that I have dedicated a whole class, `AddPostService`, with a single method, `AddPost`. Some designs create a single service class such as `PostService` and add multiple business logic methods inside it. I opted for the single public method in a single class approach to respect the single-responsibility principle of **SOLID**.

I have injected into the class two repositories that are needed for the business logic: the `author` and `post` repositories. For a reminder of DI, have a look at *Chapter 2, Understanding Dependency Injection by Example.*

I implemented a business logic that checks whether a non-existent author is passed to the method. Also, if the author is locked from publishing, then I created a post and got back the created ID. I could have used `Guid` but the UI would want an integer.

The notable thing here is that the service did not know how `Author` was loaded. It might have been loaded from a relational DB, a document DB, an in-memory DB, or even a text file! The service delegated this knowledge to the repository.

The service here focused on a single responsibility, which is the business logic for adding a new post. This is an example of the separation of concerns.

Updating title service

The title of the blog can be up to 90 characters and can be updated at any time. This is sample code to achieve this:

```
public class UpdateTitleService
{
    private readonly IPostRepository _postRepository;
    private const int TITLE_MAX_LENGTH = 90;
    public UpdateTitleService(IPostRepository postRepo)
    {
        _postRepository = postRepo;
    }
    public void UpdateTitle(int postId, string title)
    {
        if (title is null) title = string.Empty;
        title = title.Trim();
        if (title.Length > TITLE_MAX_LENGTH)
        {
            throw new
                ArgumentOutOfRangeException(nameof(title),
                $"Title max is {TITLE_MAX_LENGTH} letters");
        }
        var post = _postRepository.GetById(postId);
        if (post is null)
        {
            throw new ArgumentException(
                $"Unable to find a post of Id {postId}",
                nameof(post));
        }
        post.Title = title;
        _postRepository.Update(post);
    }
}
```

The preceding logic is straightforward. What is new here is the way the service loaded the entity, then modified one of its properties, and then asked the repository to manage the update operation.

In both services, the business logic involved no knowledge of the data platform. This can be SQL Server, Cosmos DB, MongoDB, and so on. DDD refers to the libraries for these tools as *infrastructure*, so the services have no knowledge of the infrastructure.

Application services

Earlier, we were describing the domain services. Application services provide the interaction with the outside world or the glue that allows a client to request something from your system.

A perfect example of an application service is an **ASP.NET** controller, where a controller can use domain services to provide a response to a **RESTful** request. Application services will typically use both domain services and repositories to deal with external requests.

Infrastructure services

These are used to abstract technical concerns (cloud storage, service bus, email provider, and so on).

We will be using services extensively in *Part 2, Building an Application with TDD*, of this book. So, I hope you got an idea of what they are. Later on, we will have an end-to-end project that will involve multiple services.

Service characteristics

There are guidelines on how to build a service in DDD. We will go through a few of them here. However, I recommend going through the *Further reading* section at the end of this chapter if you would like to know more.

We will discuss stateless services, ubiquitous language, and using domain objects instead of services.

Stateless

A service should not hold a state. Holding a state is akin to remembering data, which means, in plain English, persisting some business data in the fields or properties of a service class.

Avoid maintaining a state in your service as it will complicate your architecture, and if you think that you need a state, then this is what repositories are for.

Use ubiquitous language

As always, use ubiquitous language. In the previous examples, we named the services and the methods following the business operations.

Use domain objects where relevant

DDD is against anemic models, so it encourages the user to check whether a domain model can do the operation rather than having this done in a service.

In our example, DDD would have encouraged us to have behavior (public methods) in `Post`. If we were to follow the DDD advice, our `Post` class would have looked like this:

```
public class Post
{
    public int Id { get; private set; }
    public string? Title { get; private set; }
    // more properties...
    private readonly IPostRepository _postRepository;
    private const int TITLE_MAX_LENGTH = 90;
    public Post(IPostRepository postRepository)
    {
        _postRepository = postRepository;
    }

    public void UpdateTitle(string title)
    {
        ...
    }
}
```

Note that the setters for the properties are now private as only the methods within the class can set the properties. The second note is that the `UpdateTitle` method doesn't need to get `Id` as a parameter as it has access to `Id` from within the class. It only requires the new title.

The advantage of this is that your class is not anemic and follows OOP principles. Obviously, we have not followed the DDD recommendation in our implementation and wrote the `UpdateTitle` method in the service class.

I did not do this to upset DDD practitioners, but for practical purposes! Let me list the potential problems that you may encounter in this approach while using EF, the main .NET ORM:

- **DI of the repository**: You will need to instruct your DI container to inject the repository into the `Post` class at runtime. This is not a common practice, and I am not even sure whether this is possible with non-hacky code.

- **Private setters with EF**: EF cannot set your properties as they are private. So, if it loads `Post` from the database, it will be unable to set the properties, which renders EF useless.

- **Distribution of business logic**: If the domain classes contain business logic, sometimes your business logic will be in services and sometimes it will be in domain objects rather than one or the other. In other words, it will be distributed in multiple classes.

There are ways to make this work, but they aren't worth the effort. Here, practicality doesn't meet the DDD theory, and this is why I opted to have anemic domain objects. The takeaway is that you know what DDD is advocating and the reason for that and you know why we are shifting from this practice.

Services do not care how data is loaded and persisted because it is the responsibility of the repositories, which naturally leads us to the next topic.

Exploring repositories

Repositories are classes that belong to infrastructure. They understand the underlying storage platform and interact with the specifics of the data store system.

They should not contain business logic, and they should only be concerned with loading and saving data.

Repositories are a way of achieving a single responsibility (as in SOLID's single responsibility principle) by having the services and the domains responsible for business logic but not responsible for data persistence. DDD gives the data persistence responsibility to the repositories.

An example of a repository

You've seen this line of code previously in the `UpdateTitleService` class:

```
var post = _postRepository.GetById(postId);
```

Here, we will show you a potential implementation of `GetById`.

Using Dapper with SQL Server

Dapper is a .NET library categorized under the term *micro-ORM*. It is very popular and used in **StackOverflow**.

Dapper can be used to access a SQL Server DB, so assuming our blog DB is a SQL Server one, we will use Dapper to implement `GetById` of `PostRepository`.

To use Dapper in any project, you can install it via **NuGet** under the same package name. To use Dapper with SQL Server, you will also need to install the System.Data.SqlClient NuGet:

```
using Dapper;
using System.Data.SqlClient;
...
public interface IPostRepository
{
    int CreatePost(int authorId);
    Post? GetById(int postId);
    void Update(Post post);
}


public class PostRepository : IPostRepository
{
    public Post? GetById(int postId)
    {
        var connectionString = ... // Get con string from config
        using var connection = new SqlConnection
           (connectionString);
        connection.Open();
        var post = connection.Query<Post>(
            "SELECT * FROM Post WHERE Id = @Id", new {Id =
                 postId}).SingleOrDefault();
        connection.Close();
        return post;
    }
    ...
}
```

Usually, the repository classes have an interface counterpart to allow them to be injected into services. Notice that in our previous `PostService`, we have injected `IPostRepository`. The code shows how a repository works but it is not DI-compliant, however, it will be in the next section.

The `SqlConnection` class is an **ADO.NET** class, which allows you to manage a connection with a SQL Server DB.

`Query()` is an extension method provided by Dapper. It allows you to issue a regular **T-SQL** query and map the results to an object.

Using Dapper with SQL Server and DI

As you've noticed, we have not injected `SqlConnection` and we have directly instantiated it in the code. Obviously, this is not the best practice! Here is an implementation that utilizes injecting the connection object:

```
public class PostRepository : IPostRepository
{
    private readonly IDbConnection _dbConnection;

    public PostRepository(IDbConnection dbConnection)
    {
        _dbConnection = dbConnection;
    }

    public Post? GetById(int postId)
    {
        _dbConnection.Open();
        var post = _dbConnection.Query<Post>(
            "SELECT * FROM Post WHERE Id = @Id", new {Id =
                postId}).SingleOrDefault();
        _dbConnection.Close();
        return post;
    }
    ...
}
```

`SqlConnection` implements `IDbConnection` and we can wire this in the DI section in our startup to inject the right object at runtime (*not shown here, as this is a fictitious sample*). The DI will take care of instantiating the connection object, so we don't have to do it here.

The `GetById` method uses Dapper's ADO.NET extension methods to map the query results to a C# object. There are cleaner ways to achieve this, but I opted for the most readable one for this example.

Using other DBs

In the previous example, we have used a SQL Server DB; however, any other database would do. The only implementation that is going to change is within the `PostRepository` class. The consumers of `IPostRepository` will not change.

In the following chapters, we will demonstrate end-to-end implementations that use SQL Server (with EF) and Cosmos DB.

EF and repositories

EF is .NET's major ORM. An *ORM* is a term to say it loads your relational DB records into objects.

EF provides a high level of abstraction that embodies multiple DDD patterns, most notably the repositories. When using EF, the repository pattern disappears in favor of EF and the code design becomes simpler.

In this chapter, it is enough to know this. In *Chapter 9, Building an Appointment Booking App with Entity Framework and Relational DB*, we will have a complete implementation that includes EF with a fully working source code, which will clarify how things are done from end to end.

Putting everything together

This is my favorite part. I have been providing little snippets here and there and, hopefully, now you can see the big picture of how everything is linked from a DDD point of view. I have included the snippets in the source code directory.

Solution Explorer view

What we've done in this project is a collection of snippets. Let's have a look at them:

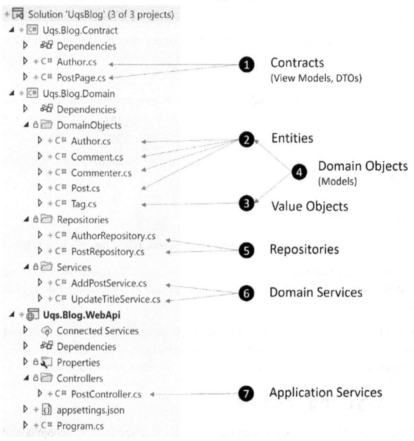

Figure 7.5 – VS solution files from a DDD perspective

Let's have a recap of every item:

1. **Contracts**: This is what the outside world sees. These contracts represent the shape of the data that will be exchanged between the backend and the client. The client should know the data elements of the contract, so it knows what to expect from your headless blog.

2. **Entities**: They are the domain objects with identities.

3. **Value Objects**: They are the domain objects that don't require an identity.

4. **Domain Objects**: This is the group of entities and value objects in your system.

5. **Repositories**: These are the classes that will save and load your data from a data store (relational DB, document DB, file system, blog storage, and so on).

6. **Domain Services**: This is where the business logic will live, and it will interact with the repositories for CRUD operations. These services are not exposed to the outside world.

7. **Application Services**: Controllers in basic scenarios act as application services where they interact with domain services to serve a REST request. Application services are exposed to the outside world.

It also happened that we only have a single aggregate, which is all our domain objects. A domain might have more than one aggregate. We also have Post as our aggregate root.

Architectural view

We've seen a potential project and file structure for our DDD project and now, let's have a look at it from a software design point of view:

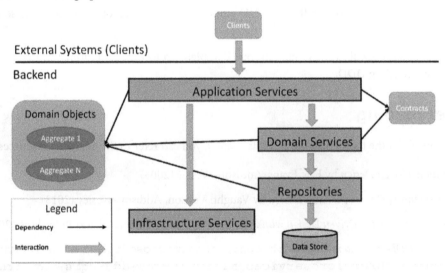

Figure 7.6 – A simplified software design view of DDD

Let's discuss this DDD-style system:

- **Application Services**: They interact with the clients and the domain services. They deliver the data to a client based on the contract and they deal with domain services directly.

- **Domain Services**: They provide services to **Application Services**.

- **Infrastructure Services**: They provide services that are not part of the domain, such as fetching the ZIP code/postcode city.

- **Aggregate**: Each aggregate contains several domain objects and has one aggregate root.

- **Domain Objects**: They are all the entities and value objects in all aggregates.

I hope I was able to show you the foundation of the DDD design from the coding and projects structure and from an architectural view as well, although risking repeating the concepts twice.

Summary

There are topics in DDD that I have omitted, as they don't contribute directly to the rest of the book, such as bounded contexts, domain events, units of work, and others. I have provided additional resources in the *Further reading* section that will help you to explore the concepts further.

We have discussed the basics of DDD and I am expecting this chapter to make you familiar with this concept, so we can use the terms such as *domain objects*, *domain services*, and *repositories* freely in later chapters without you raising an eyebrow. We have also seen sample code of the different constituents of DDD.

We have also seen where we will shift from DDD guidelines where it is more practical to do so and explained why.

In the next chapter, we will set a foundation for a complete project that will utilize all that you've learned so far, including DDD.

Further reading

To learn more about the topics discussed in this chapter, you can refer to the following resources:

- *Domain-Driven Design* by Eric Evans, Addison-Wesley (2003)

- *Implementing Domain-Driven Design* by Vaughn Vernon, Addison-Wesley (2013)

- *Hands-On Domain-Driven Design with .NET Core* by Alexey Zimarev, Packt Publishing (2019)

- *Design a DDD-oriented microservice*: `https://docs.microsoft.com/en-us/dotnet/architecture/microservices/microservice-ddd-cqrs-patterns/ddd-oriented-microservice`

- *Martin Fowler on DDD*: `https://martinfowler.com/bliki/DomainDrivenDesign.html`

- *Quickstart: Build a console app by using the .NET V4 SDK to manage Azure Cosmos DB SQL API account resources*: `https://docs.microsoft.com/en-us/azure/cosmos-db/sql/create-sql-api-dotnet-v4`

- *Dapper on GitHub*: `https://github.com/DapperLib/Dapper`

8

Designing an Appointment Booking App

In previous chapters, we've seen sample implementations that were limited in scope because it would be impractical to have a full application on every covered topic.

This chapter covers the design of a barber appointment booking application, which will combine what we've learned from previous chapters:

- Dependency injection
- Unit testing
- Test doubles using mocks and fakes
- DDD
- Applying TDD

Chapter 9 and *10* will cover the implementation of this chapter. This chapter is about the business requirements and design decisions, not about the implementation (the code).

Before proceeding with this chapter and the rest of *Part 2*, I would highly recommend that you are familiar with the topics that I've listed above. They are all covered in *Chapter 2* to *Chapter 7*.

In this chapter, we will cover the following:

- Business requirements to build a booking system
- The design of the system DDD-style
- The implementation routes of this system

By the end of this chapter, you will understand better a realistic DDD analysis based on a life-like problem.

Technical requirements

The code for this chapter can be found in the following GitHub repository:

```
https://github.com/PacktPublishing/Pragmatic-Test-Driven-Development-
in-C-Sharp-and-.NET/tree/main/ch08
```

Collecting business requirements

You work for a software consultancy called **Unicorn Quality Solutions** (**UQS**), which is implementing an appointment booking application for Heads Up Barbers, a modern barber shop with many employees.

The required application will comprise three applications:

- **Appointment booking website**: Where customers will book an appointment for hairdressing.

- **Appointment booking mobile app**: Same as the website, but a native mobile app (as opposed to a website on a mobile web browser).

- **Back office website**: This is an internal app to be used by the owner of the business. It allocates shifts for barbers (employees), cancels bookings, calculates the barbers' commission, and so on.

Phase 1 of the delivery is only the first application (booking website), which has the highest business value because it allows the users to book via desktop and on their mobile web browser.

This is our concern for the rest of Part 2 of this book. The following is a diagram showing the three phases of the project:

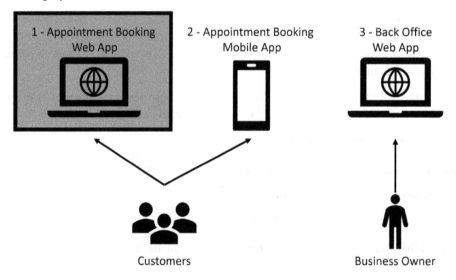

Figure 8.1 – The three required applications

Although we are only concerned with building *Phase 1*, we need to consider in our design that our architecture will include support for a mobile app at later phases.

Business goals

In this day and age, most customers like to book an appointment online, especially since COVID-19, where shops tried to reduce the concentration of people in spaces via appointments.

Heads Up Barbers wants a booking solution that aims to do the following:

- Market the available hairdressing services.
- Allow a customer to book an appointment with a specific or a random barber.
- Give barbers a rest between appointments, usually 5 minutes.
- Barbers have various shifts in the shop and they are off work on different days, so the solution should take care of picking free slots based on the availability of barbers.
- Time saving by not having to arrange appointments on the phone or in person.

Stories

After analyzing the business goals, UQS came up with more detailed requirements in the form of user stories and mockups. We will go through these next.

Story 1 – services selection

As a customer:

I want to have a list of all available services and their cost.

So I can select one for booking.

And be transferred to the booking page.

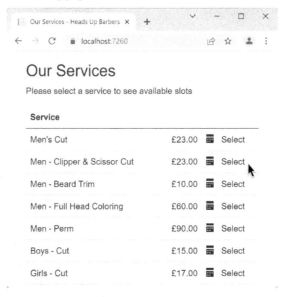

Figure 8.2 – The list of the available services and their prices

This mockup displays all the available services with their prices and **Select** hyperlinks to take the user to the booking page for the selected service.

Story 2 – default options

As a customer:

I want to have a booking page with [**Any employee**] and today's date selected by default.

So I spend less time clicking and finish booking faster.

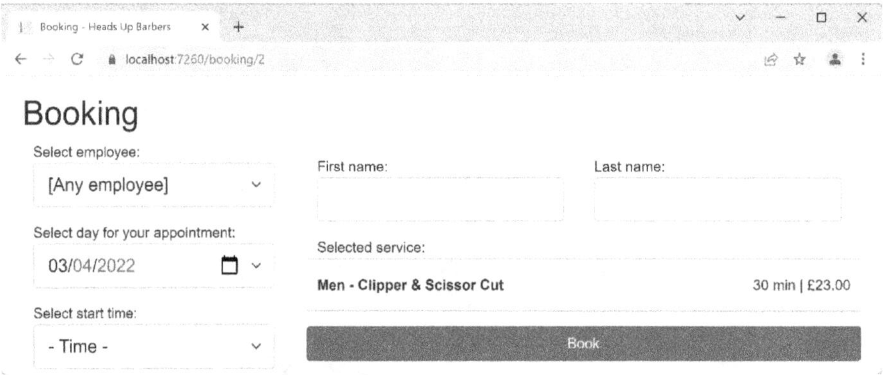

Figure 8.3 – The booking page with default options already selected

Notice that [**Any employee**] and the current day, **2022-04-03**, are selected by default.

Story 3 – select employee

As a customer:

I want to select any employee or a specific employee for my appointment.

So I can pick my favorite barber if I have one.

Figure 8.4 – Selecting a specific employee

The customer will have a list of barbers working for Heads Up Barbers from which they can pick their favorite one.

Story 4 – appointment days

As a business:

We want to present the customer with a 7-day window max, including the current day, to pick an appointment.

And we want to reduce this window if the selected employee is not fully available.

So we can guarantee our employees' availability for booking.

Figure 8.5 – Calendar showing a 7-day window starting 2022-04-03

The mockup will take into consideration the changes in the selected employee's schedule and show only the availability window for the selected employee.

Story 5 – time selection

As a business:

I want to present the customer with the time slots available for the selected employee for the selected date.

And take into consideration existing employee appointments and the employee's shifts.

And round up any appointment to the nearest 5 minutes.

And take into consideration the rest time of 5 minutes between appointments.

So I ensure the customer is selecting an employee that is already available.

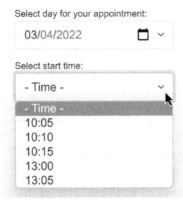

Figure 8.6 – Time slots available for the employee for the selected date

Let's take a few examples to clarify the requirements.

Notice that all the minutes are multiples of 5.

Example 1 – no shifts are available

If an employee has no allocated shifts on the selected date, the list will be empty and the customer will be unable to book.

Example 2 – no appointments are booked

An employee, Tom, has a shift on 2022-10-03 from 9:00 to 11:10 and has no booked appointments. The customer wants to book a 30-minute-long service. The selected start time will have the following values: 09:00, 09:05, 09:10, …, 10:35, and 10:40.

Example 3 – multiple appointments booked at the end of the shift

An employee, Tom, has a shift on 2022-10-03 from 9:00 to 11:10, but he already has appointments booked from 09:35 to 11:10. The customer wants to book a 30-minute-long service. The selected start time will have the following value: 09:00. The following figure illustrates the time spans:

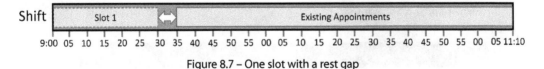

Figure 8.7 – One slot with a rest gap

Example 4 – multiple appointments booked at the end of the shift

Tom has a shift on 2022-10-03 from 9:00 to 11:10, but he already has appointments booked from 09:40 to 11:10. The customer wants to book a 30-minute-long service. The selected start time will have the following values: 09:00 and 09:05.

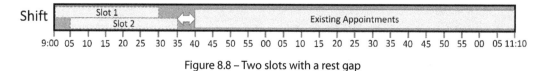

Figure 8.8 – Two slots with a rest gap

Example 5 – an appointment booked in the middle of the shift

Tom has a shift on 2022-10-03 from 9:00 to 11:10, but he already has appointments booked from 09:40 to 10:35. The customer wants to book a 30-minute-long service. The selected start time will have the following values: 09:00, 09:05, and 10:40.

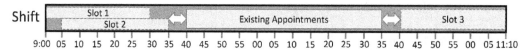

Figure 8.9 – Three slots with two rest gaps

Story 6 – name filling

As a customer:

I have to fill in my first and last name to act as my ID when I show up at the barber shop.

So I am uniquely identified.

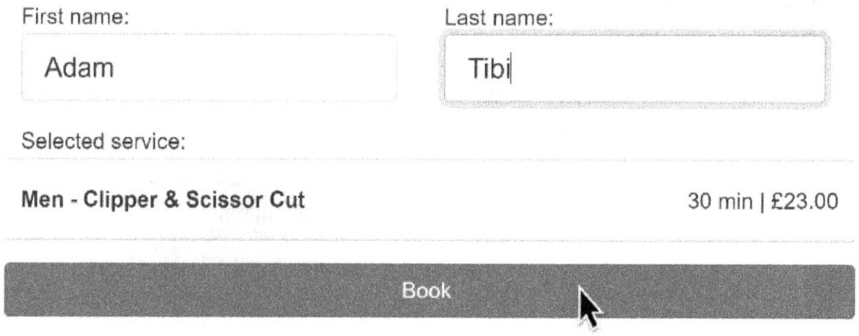

First name:

Adam

Last name:

Tibi

Selected service:

Men - Clipper & Scissor Cut   30 min | £23.00

Book

Figure 8.10 – First name and last name fields

Story 7 – service display

As a customer:

I want a reminder of the name of the service that I picked, its price, and the required time.

So I can review my selection before hitting the **Book** button.

Story 8 – all fields are mandatory validation

As a customer:

I have to select and fill in all fields before booking.

So I won't get validation errors.

Story 9 – random selection with any employee

As a business:

When [**Any employee**] is selected.

And more than one employee is free at the selected slot.

And I hit **Book.**

A free employee is selected *randomly.*

So I ensure our employees are allocated to appointments fairly.

Example 1 – three employees are free at one slot

If the customer selects [**Any employee**] and gets three employees (Thomas, Jane, and William) who are free at 09:00, and the customer selects **09:00** and hits **Book**, Thomas, Jane, or William is allocated randomly to the appointment without taking into consideration any other factor, and one of them is selected.

Story 10 – confirmation page

As a customer:

I want to see that my appointment is booked.

So I can rest assured that it is going ahead.

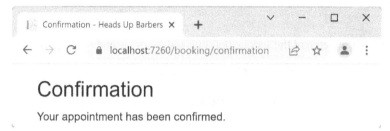

Figure 8.11 – Confirmation page

The confirmation page above is a simple static page.

You probably sensed that *Story 5* is the most demanding one from a business logic perspective, and this will be heavily targeted by our unit tests.

As you can see, the scope of the implementation is limited. In the future, we can extend this further with:

- Online payment
- User login
- Email confirmation
- And more…

However, the stories so far describe a robust life-like system. Some might call this a **minimum viable product** (**MVP**); however, I wouldn't as it might wrongly imply a lower-quality system.

Now it's time to move from the business requirements to the general guidelines for designing our system.

Designing with the DDD spirit

We have learned in the previous chapter an overview of DDD. In our implementation, we will follow the spirit of DDD to design the business classes.

Domain objects

If we were to read all the stories and think of a domain model, we might come up with the following classes:

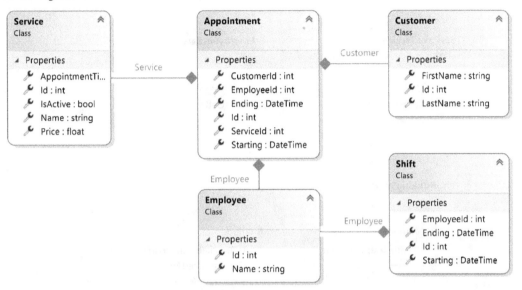

Figure 8.12 – A diagram of the domain classes

- **Service**: Represents the service offered by the barber, with `AppointmentTimeSpanInMin` being the duration of the service and `IsActive` being true to offer it to the client.

- **Customer**: Represents a customer. We are currently only interested in their name.

- **Employee**: This class will expand at a later phase to have more info, but for now, we only need the name.

- **Shift**: Represents a unique availability time for the barber. The back office application (not within scope) will allow the business owner to add shifts for employees on a daily basis to cover at least 7 days forward. So, whenever we present the customer with days selection, we have at least 7 days in the future.

- **Appointment**: It is clear that an appointment links a service to an employee and a customer. It also specifies the beginning and end times of the appointment.

We have a single *aggregate* in our implementation with all the previous classes, and our *aggregate root* is clearly the `Appointment` class.

Domain services

Domain services contain the business logic that governs the system behavior. Our system will be dealing with four categories of business logic, which could lead to four domain services:

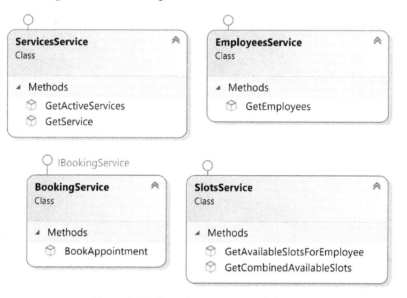

Figure 8.13 – Domain services initial design

The services at this stage are just an initial design. You usually design services driven by a TDD process rather than designing the services in advance, and this is usually done in serial, one service after the other.

System architecture

While we are only doing *Phase 1* of the system, our architecture should be ready for future phases given that a mobile app, which uses the same logic as the booking website, will be implemented in the next phase. With this in mind, the architecture in the next diagram can support all phases:

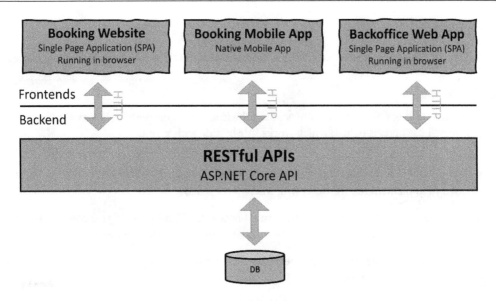

Figure 8.14 – Architecture design

Having one backend to support all the clients would embed one business logic to support all clients, so all our business logic will be behind our RESTful API application.

Also, this would make our backend act as a monolith application made up of a collection of APIs in one project and a single DB. This is alright, as this is a project with a limited scope, and going down the microservices route would be overkill.

This is a well-known architectural model where you hide the business logic behind web APIs to support multiple clients and make the logic centralized. There should be no restructuring of the architecture with the future phases when we add the booking mobile app and the back office web app.

Implementation routes

We are going to implement the backend in different ways. Each implementation will yield the same API outcome, but the point of this would be experiencing multiple unit tests and test double scenarios with each implementation.

Your team might be using one of these architectural routes, as they might be utilizing a document DB or a relational DB as in the case of most modern apps.

Frontend

In this book, we focus more on the backend, so, implementing TDD on the frontend is not covered.

> **Important Note**
>
> There are unit testing frameworks that would test the front-end. One popular library for Blazor, which we will use here, is **bUnit,** which works side by side with xUnit.

Among all popular JavaScript **single page application (SPA)** platforms, such as React, Angular, and Vue, I decided to implement the frontend with Microsoft's **Blazor**.

Blazor is a web framework that relies on C# instead of JavaScript. Simply put, Blazor converts C# into a low-level language called **WebAssembly (Wasm)** that is understood by the browser.

I chose Blazor as I am assuming it would be easier for a C# developer without SPA experience or JavaScript/TypeScript experience.

The implementation for the frontend is minimal, and the preceding mockup screenshots in the *Stories* section are taken from the Blazor application. You can find it in this chapter's GitHub under Uqs. AppointmentBooking.Website.

> **Important Note**
>
> The implementation of this frontend is aimed at readability and minimalism, rather than web design, UX, robustness, and best practices.

To launch the website:

1. Open UqsAppointmentBooking.sln in VS.
2. Right-click on Uqs.AppointmentBooking.Website and select **Set as a Startup Project**.
3. Run from VS.

Feel free to run the website and click around. You will note that it is mocked, so it is not relying on a real DB but on sample data. The discussion about the frontend is limited to this section, as the focus of the book is TDD and the backend.

Relational database backend

Usually, using a relational DB such as SQL Server and Oracle invites **Entity Framework (EF)**. Having your backend relying on EF has an effect on the way you organize your tests and the test double types that you are going to use.

Chapter 9, *Building an Appointment Booking App with Entity Framework and Relational DB*, will be dedicated to implementing the requirements with a relational database (SQL Server) and with EF.

Document DB backend

When using a document DB such as Cosmos DB, DynamoDB, and MongoDB, you do not use EF. That means you will be implementing more DDD patterns such as the *Repository pattern*. This will make the implementation with a document DB fairly different than the one that uses EF from a test doubles and **dependency injection (DI)** point of view.

Chapter 10, *Building an App with Repositories and Document DB*, will be repeating the implementation of *Chapter 9* but with around 50% different code, as it will be using a document DB.

Presenting both versions will allow you to see the difference between the implementations and, hopefully, promote your understanding of test doubles and DI. However, if you are only interested in a particular type of DB, then you can choose *Chapter 9* or *Chapter 10*.

The good news is that there are repetitions between these two chapters, where you will be able to spot them easily and focus on the unique implementation.

Using the Mediator pattern

When using the Mediator pattern, all your design changes, and your testing and test doubles follow suit. The Mediator pattern is a two-edged sword; it has a steep learning curve, but when learned and implemented, it provides a higher level of component separation of concern. It will also alter the structure of your unit tests. The Mediator pattern is outside the scope of this book, and it is mentioned here to point you to discover related patterns that affect your DI implementation and your unit tests.

Hopefully, by the end of *Part 2*, you've got a real sense of how to implement TDD in a more realistic setting.

Summary

We've seen fairly decent user requirements and we've seen a potential design for the system. This chapter was the beginning of *putting everything together*.

You've also seen a design based on DDD, which will change into code in later chapters. We have also discussed implementation routes that will affect the way we do testing and test doubles.

Sophisticated and modern projects use concepts from DDD. By now, after analyzing a full project, I hope that the DDD terminology will start sounding familiar and aid you in building your next project and help you communicate with expert developers.

The next chapter is an implementation of this chapter, but with a focus on SQL Server and EF.

Further reading

To learn more about the topics discussed in the chapter, you can refer to the following link:

- *Mediator NuGet popular lib in .NET*: `https://github.com/jbogard/MediatR`

Building an Appointment Booking App with Entity Framework and Relational DB

In the previous chapter, we outlined the technical specifications and design decisions for building an appointment booking system for a barber's salon called Heads Up Barbers. This chapter is a continuation of *Chapter 8, Designing an Appointment Booking App*, so I strongly advise you to be familiar with what was covered in that chapter first.

This chapter will implement the requirements in TDD style and will use **Entity Framework (EF)** and SQL Server. The implementation will be applicable to other **Relational Database Management Systems (RDBMSs)** such as Oracle DB, MySQL, PostgreSQL, and others.

If you are a fan of relational DBs or you are using one at work, then this chapter is for you, whereas if you are using a document database, then you might want to skip this chapter and go to the next one. Both chapters, *Chapter 9* and *Chapter 10*, have the same outcome, but they use different types of backend databases.

I assume you are familiar with EF and how it is wired and used. However, if you are not, then I encourage you to familiarize yourself with it first.

In this chapter, we will cover:

- Planning the code and the project structure
- Implementing the WebApis with TDD
- Answering frequently asked questions

By the end of the chapter, you will have experienced the implementation of an end-to-end app using TDD with mocks and fakes. Also, you will witness the analysis process that precedes writing unit tests.

Technical requirements

The code for this chapter can be found in the following GitHub repository:

https://github.com/PacktPublishing/Pragmatic-Test-Driven-Development-in-C-Sharp-and-.NET/tree/main/ch09

To run the project, you will need to have a flavor of SQL Server installed. This can be, Azure SQL, SQL Server Express LocalDB or any other SQL Server flavor.

The implementation doesn't use any advanced SQL Server features, so feel free to use any. I have tested the application with SQL Server Express LocalDB. You can find more about it here:

https://docs.microsoft.com/en-us/sql/database-engine/configure-windows/sql-server-express-localdb

You can also use any other RDBMS, but you will have to change the DB provider in the code to use the specific .NET DB provider.

To run the project, you have to modify the connection string to your specific DB instance in Uqs.AppointmentBooking.WebApi/AppSettings.json. Currently, it is set to:

```
"ConnectionStrings": {
    "AppointmentBooking": "Data
      Source=(localdb)\\ProjectModels;Initial
      Catalog=AppointmentBooking;Integrated Security=True;…"
},
```

The connection string is pointing to LocalMachine and will connect to a database called AppointmentBooking.

If you decided to use an alternative RDBMS, then you will have to install the relevant NuGet packages in Uqs.AppointmentBooking.WebApi and change the following code in Program.cs of the same project to your specific RDBMS:

```
builder.Services
    .AddDbContext<ApplicationContext>(options =>
    options.UseSqlServer(
        builder.Configuration
            .GetConnectionString("AppointmentBooking")
    ));
```

The previous DB configuration steps are optional. You can implement the requirements for this chapter without using a DB, but you won't be able to run the project and interact with it in the browser.

Planning your code and project structure

In *Chapter 8*, *Designing an Appointment Booking App*, we planned our domain and analyzed what we needed to do. The project architecture will follow the classical three-tier applications of the client application (the website), business logic (the web APIs), and database (SQL Server). Let's translate this into VS solutions and projects.

In this section, we will create the solution, create the projects, and wire up the components.

Analyzing the project's structure

Ask a group of senior developers to come up with a project structure, and you will end up with multiple structures! In this section, we will discuss a way of organizing your project structure that I have developed over the years.

Given that we are first going to build a website for the user and later a mobile app (not covered in this book), it makes sense to isolate the business logic to a WebApi project that can be shared by both the website and the mobile app. So, we will build a website project based on Blazor WebAssembly called `Uqs.AppointmentBooking.Website`.

The domain logic will be exposed as APIs, so we will create an ASP.NET API project for the APIs called `Uqs.AppointmentBooking.WebApi`.

The previous two projects need to exchange data in an agreed structure called **Data Transportation Objects (DTOs)**, commonly known as *contracts*, so, we will create a .NET library project called `Uqs.AppointmentBooking.Contracts`. This project will be referenced by both the website and the WebApi projects.

The WebApi project translates web requests into something we can understand in C#. In technical terms, this will manage the HTTP communication layer with RESTful-style APIs. So, the WebApi project will not contain business logic. The business logic will be in our domain project. We will create a domain project called `Uqs.AppointmentBooking.Domain`.

Your business logic will live in two places – the UI and the domain layer. The UI business logic will manage UI functionalities such as toggling dropdowns, blocking calendar days, responding to drag and drop, and disabling/enabling buttons, among others. This logic will live in the website project. The language used in writing the code depends on the UI framework in use, such as Angular, React, and Blazor. Usually, you do not implement the features of the UI project using TDD, but you can use unit tests. In our implementation, we will have little code in the UI layer, so we will not be doing any UI unit tests.

The sophisticated business logic will live in the domain layer, and we will be writing it following the concepts of TDD. So, we shall create a project that will hold our domain unit tests and call it `Uqs.AppointmentBooking.Domain.Tests.Unit`.

To put these projects into perspective and map them to our 3-tier architecture, we can have the following diagram:

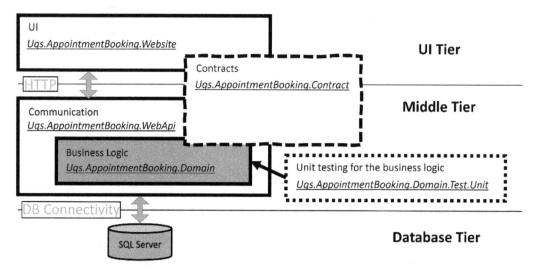

Figure 9.1 – The relationship between the projects and the application design

The previous diagram shows the functionality that each project provides to form the 3-tier application. Let's start by creating the VS solution structure.

Creating projects and configuring dependencies

This is the unavoidable boring part, creating the solution and projects and linking them together. In the following section, we will take the command-line approach rather than the UI approach.

> **Note**
>
> I have added a text file called create-projects.bat to the project source control that contains all the command lines, so you don't have to manually write them. You can copy and paste this file to your desired directory and then, from your command line, execute the file.

The following is a list of the commands required to create your VS solution and its projects:

- From your OS console, navigate to the directory where you want to create your new solution and execute the following to create the solution file:

```
md UqsAppointmentBooking
cd UqsAppointmentBooking
dotnet new sln
```

- Execute this to create the projects and notice that we are using a different template for each project:

```
dotnet new blazorwasm -n Uqs.AppointmentBooking.Website
dotnet new webapi -n Uqs.AppointmentBooking.WebApi
dotnet new classlib -n Uqs.AppointmentBooking.Contract
dotnet new classlib -n Uqs.AppointmentBooking.Domain
dotnet new xunit -n
   Uqs.AppointmentBooking.Domain.Tests.Unit
```

- Add the projects to the solution:

```
dotnet sln add Uqs.AppointmentBooking.Website
dotnet sln add Uqs.AppointmentBooking.WebApi
dotnet sln add Uqs.AppointmentBooking.Contract
dotnet sln add Uqs.AppointmentBooking.Domain
dotnet sln add Uqs.AppointmentBooking.Domain.Tests.Unit
```

- Now let's set up dependencies between the projects:

```
dotnet add Uqs.AppointmentBooking.Website reference
   Uqs.AppointmentBooking.Contract
dotnet add Uqs.AppointmentBooking.WebApi reference
   Uqs.AppointmentBooking.Contract
dotnet add Uqs.AppointmentBooking.Domain reference
   Uqs.AppointmentBooking.Contract
dotnet add Uqs.AppointmentBooking.WebApi reference
   Uqs.AppointmentBooking.Domain
dotnet add Uqs.AppointmentBooking.Domain.Tests.Unit
   reference Uqs.AppointmentBooking.Domain
```

And the last bit is adding the required NuGet packages to the project. The domain project will communicate with the SQL Server database using EF. The `Microsoft. EntityFrameworkCore.SqlServer` package allows the required libraries to connect the project to SQL Server. To add this library to the `Domain` project, use the following:

```
dotnet add Uqs.AppointmentBooking.Domain package
   Microsoft.EntityFrameworkCore.SqlServer
```

- The unit testing project will require *NSubstitute* for mocking, so let's add its NuGet:

```
dotnet add Uqs.AppointmentBooking.Domain.Tests.Unit
    package NSubstitute
```

- We will be using a fake to test double EF. This fake will create an in-memory database that will make our testing easier to write. We will discuss this in more detail later in this chapter, but for now, let's add this fake library:

```
dotnet add Uqs.AppointmentBooking.Domain.Tests.Unit
    package Microsoft.EntityFrameworkCore.InMemory
```

For visual inspection, you can open the solution file with VS, and it should look as such:

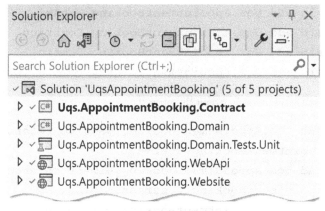

Figure 9.2 – A view of VS solution explorer

At this stage, your solution structure should look similar.

Now that the project structure is in place, we will modify the code.

Setting up the domain project

From the domain analysis in *Chapter 8, Designing an Appointment Booking App*, we have created a list of the domain objects. I will not go through them again; I will just create and add them to the Domain project under DomainObjects:

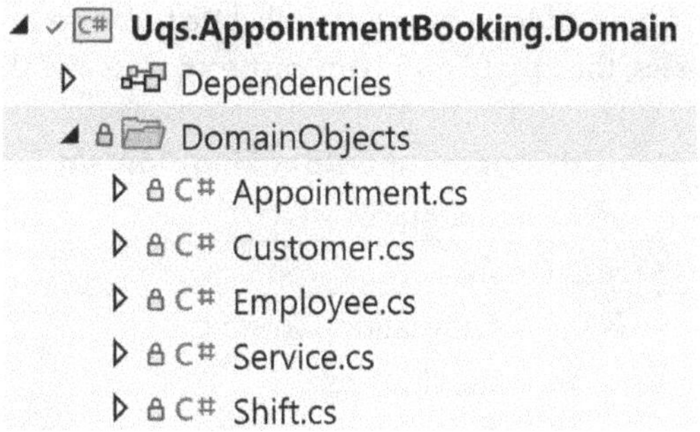

▲ ✓ C# **Uqs.AppointmentBooking.Domain**
  ▷   🖧 Dependencies
  ▲ 🔒📁 DomainObjects
    ▷ 🔒 C# Appointment.cs
    ▷ 🔒 C# Customer.cs
    ▷ 🔒 C# Employee.cs
    ▷ 🔒 C# Service.cs
    ▷ 🔒 C# Shift.cs

Figure 9.3 – Added domain objects

These are just data structures with no business logic. Here is the source code of one of them, the
`Customer` domain object:

```
namespace Uqs.AppointmentBooking.Domain.DomainObjects;
public class Customer
{
    public int Id { get; set; }
    public string? FirstName { get; set; }
    public string? LastName { get; set; }
}
```

You can view the rest of the files in the chapter's GitHub repo online.

Next is wiring up the focus of this chapter, EF.

Wiring up Entity Framework

We are going to use EF to store each domain object in a database table that bears the same name
but in the plural, as this is the default behavior of EF. So, the `Customer` domain object will have a
Customers table equivalent in the DB.

We will not be customizing much in EF as our intention in this chapter is to focus on TDD, as doing
the little setups here and there are only chores and you can find them in the companion code.

Under the `Domain` project, I have added a directory called `Database` to contain our EF-related classes. We will need two classes, the `ApplicationContext` class and the `SeedData` class:

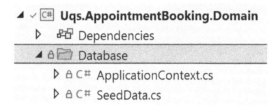

Figure 9.4 – Added EF files

In the next section, we will discuss their role.

Adding the context class

With EF, you add a context class to reference all your domain objects. I called my context class `ApplicationContext`, and I followed basic EF practices. Here is my class:

```
public class ApplicationContext : DbContext
{
    public ApplicationContext(
      DbContextOptions<ApplicationContext> options) :
      base(options){}
    public DbSet<Appointment>? Appointments { get; set; }
    public DbSet<Customer>? Customers { get; set; }
    public DbSet<Employee>? Employees { get; set; }
    public DbSet<Service>? Services { get; set; }
    public DbSet<Shift>? Shifts { get; set; }
}
```

This is the most basic setup of EF with no customization, with every property mapped to a database table name.

From this point onward, we will use `ApplicationContext` to execute operations on the DB.

Let's continue with our process and set up EF within WebApi.

Wiring up EF with the WebApi project

The WebApi will wire EF to the right DB provider, which, in this case, is SQL Server, and will pass the connection string to EF at runtime.

So, the first step is to add the connection string to the WebApi's `AppSettings.js`:

```
"ConnectionStrings": {
  "AppointmentBooking": "Data
    Source=(localdb)\\ProjectModels;Initial
    Catalog=AppointmentBooking; (...) "
},
```

Obviously, the connection string may vary based on your DB location and setup.

> **Note**
>
> In this chapter, I am not concerned with setting multiple environments, but you may want to create multiple `AppSettings` for different environments and change the connection string accordingly.

The next step is to wire up the WebApi with EF and provide it with the connection string. This should be done in `Program.cs`, preferably directly after the first line, `var CreateBuilder(args)`:

```
var builder = WebApplication.CreateBuilder(args);
builder.Services.AddDbContext<ApplicationContext>(o =>
  o.UseSqlServer(builder.Configuration.GetConnectionString
    ("AppointmentBooking")));
```

This is what we need for wiring up EF. However, for development purposes, we might need some test data to fill the pages with some meaningful data. We shall do this next.

Adding seed data

Newly created DBs have empty tables, and the `seed` class is meant to pre-populate the tables with sample data.

I will not list the code here as it is outside the scope of the chapter, but you can look at the code in the `Domain` project in `Database/SeedData.cs`.

We've just finished the setup for the WebApi project, which is going to be consumed by the website, so let's create the website next.

Setting up the website project

Phase one of this implementation includes creating a website to access the APIs to provide a UI for the user, which we did previously in this chapter by command line. However, website implementation is outside the scope of this chapter and the book in general, as it is not related to TDD, so I will not be going through the code.

Though, we are interested in one aspect – what does the website require from the web apis? We will need to understand this in order to build the required functionality in `WebApis` the TDD way.

We will answer this question bit by bit in the next section of this chapter.

In this section, we covered the setup and configuration aspect of the project, and we have not done anything that is affected by TDD. You may have noticed that I referred you to the companion source code on multiple occasions; otherwise, we would have no place left for the core of this chapter, the TDD part, which we will do next.

Implementing the WebApis with TDD

To build the WebApi project, we are going to look at each requirement from *Chapter 8*, *Designing an Appointment Booking App*, and provide the implementation that satisfies it using TDD style.

The requirements are all stated in terms of the Website and its functionality, and they do not dictate how to build our APIs. The Website will have to call the WebApis for any business logic as it has no access to the DB and deals with UI-related business logic only.

This chapter is dedicated to EF for a good reason as we want you to appreciate *fakes*, which are not as popular as *mocks*, both from the test doubles family. Also, it will be a typical example of a .NET solution of an ASP.NET Core and a relational DB implementation.

In this section, we will cover working in TDD mode, taking into consideration our persistence provider, EF.

Using the EF in-memory provider

To make our life easier when unit testing the system, we want to abstract the database in an elegant way. When I say elegant, I mean less code and more readability.

However, the challenge we face when testing a system that has a DB is that we do not want to hit the real DB in our unit tests as this would defeat the whole purpose of unit testing and make it a sort of integration or Sintegration testing. Instead, we use test doubles to abstract it. A fake is a test double that replaces a component during unit testing with an equivalent component more suitable for testing purposes, we will be employing a fake to replace our DB for unit testing purposes.

EF has a provider that accesses SQL server, which is what we want to use in production during the system run, but in unit testing, we can't do this. Lucky for us, EF has what is called an in-memory provider, which can build and destroy an in-memory database during each unit test run.

Building and destroying an in-memory DB during unit testing is cheap compared to doing the same for a physical DB, not to mention the cost and the possibility of random errors that are generated from trying to create and delete a real database so often (with each single unit test execution). As you might have already figured out, the EF in-memory provider acts as a fake.

During runtime, we use the SQL Server provider, and during unit testing, we use the in-memory provider, and we accomplish this switch via dependency injection:

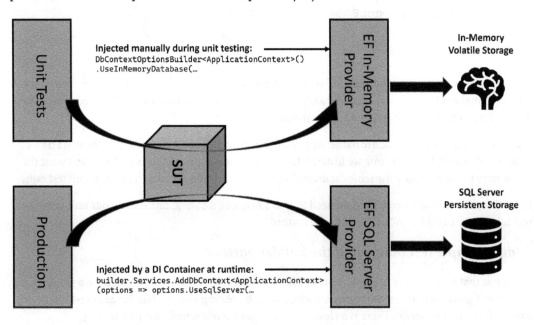

Figure 9.5 – Runtime and test times with respect to EF providers

The previous diagram illustrates injecting different providers in different project stages. The unit testing stage will use the EF in-memory provider and in the production run stage, the proper production provider, EF SQL Server Provider, will be used.

Configuring the in-memory provider

To get the advantage of the in-memory provider, I have created a file in the unit tests project called `ApplicationContextFake.cs`, and here is the code:

```
public class ApplicationContextFake : ApplicationContext
{
    public ApplicationContextFake() : base(new
        DbContextOptionsBuilder<ApplicationContext>()
        .UseInMemoryDatabase(databaseName:
        $"AppointmentBookingTest-{Guid.NewGuid()}")
        .Options) {}
}
```

Note that we are inheriting the main EF object, `ApplicationContext`, and we configured the option to make it in-memory. `ApplicationContextFake` is meant to be injected whenever `ApplicationContext` is required in our unit tests.

We are creating a unique database name, `AppointmentBookingTest-{Guid.NewGuid()}`, by appending a GUID every time we instantiate the fake. The reason for this is that we don't want the in-memory provider to have the same database name to avoid caching any data in between unit test calls.

From this point onward, every time we need to inject `ApplicationContext` in our unit tests, we will inject `ApplicationContextFake` instead.

Adding sample test data using the builder pattern

Every test that we are going to implement will have a state. For example, we might have a single free barber or a group of barbers with different schedules, so creating sample data for each test might be a mess if we are not careful. There is a clever way to organize our sample data for testing.

We can do this in a pattern called the builder pattern (not to be confused with the GoF Builder design pattern). The builder pattern will allow us to *mix and match* sample data in a clean and readable way. I have added a file called `ApplicationContextFakeBuilder.cs` to contain the sample state data with the builder pattern. I have included a part of this class here for brevity but you can see the full class in the companion source code:

```
public class ApplicationContextFakeBuilder
{
    private readonly ApplicationContextFake _ctx = new();
    private EntityEntry<Employee> _tomEmp;
    private EntityEntry<Employee> _janeEmp;
```

```
    ...
    private EntityEntry<Customer> _paulCust;
    private EntityEntry<Service> _mensCut;
    private EntityEntry<Appointment> _aptPaulWithTom;
    ...
    public ApplicationContextFakeBuilder WithSingleEmpTom()
    {
        _tomEmp = _ctx.Add(new Employee {
            Name = "Thomas Fringe" });
        return this;
    }
    ...
    public ApplicationContextFake Build()
    {
        _ctx.SaveChanges();
        return _ctx;
    }
}
```

This class will prepare in-memory sample data. The unit tests that will use this class will call different methods on it to set the right data state. What is interesting in this class is the following:

- Using the `With` convention to indicate that we are adding sample data. You will see examples of how the `With` methods are used later on.

- The `With` methods return `this`, which at first instance looks weird. The idea here is to implement a coding convention called chaining so that you can write code like this: `_ctxBldr.WithSingleService(30).WithSingleEmpTom()`.

- The `Build()` method will save everything to the persisting media (the memory, in this case) and return the context.

The Builder pattern is heavily used when trying to set the state of a certain component. Feel free to look at the companion code for the full code. *Chapter 6, The FIRSTHAND Guidelines of TDD*, has another example of a build class; you might want to have a look at it to strengthen your understanding.

Implementing the first story

The first story in our requirement is very easy. The website is going to display all the available services that we have. Since the website will request this data from the WebApi through a RESTful API call, the domain layer will have a service that will return this list. Let's assume this would be the UI output:

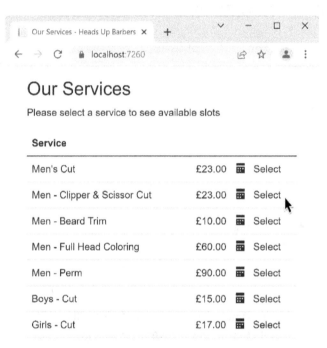

Figure 9.6 – A UI of the requirements of Story 1

The UI layer, hosted in the browser, will need to issue a RESTful call to the WebApi, which can look as follows:

```
GET https://webapidomain/services
```

This UI will require a few data properties that should be returned by this API. So, the fetched JSON can look like an array of this:

```
{
    "ServiceId": 2,
    "Name": "Men - Clipper & Scissor Cut",
    "Duration": 30,
    "Price": 23.0
}
```

You can see where each part is used on the page, but maybe `ServiceId` is not very clear. It will be used to construct the URL of the `select` hyperlink. So, we can now design the contract type that will render this JSON, which could look like this:

```
namespace Uqs.AppointmentBooking.Contract;
public record Service(int ServiceId, string Name,
    int Duration, float Price);
```

This `record` contract will render the previous JSON code. And the full returned array contract could look like this:

```
namespace Uqs.AppointmentBooking.Contract;
public record AvailableServices(Service[] Services);
```

You can find these contract types and all the other contracts in the `Contract` project.

Adding the first unit test via TDD

Thinking along the lines of DDD, we will have a domain service called `ServicesService`, which will handle retrieving all the available services. So, let's look at the structure of this service. We will create it in the `Domain` project under `Services`. Here is the code:

```
public class ServicesService
{
}
```

There is nothing special here. I have just helped VS understand that when I type `ServicesService`, it should guide me to this class.

> **Note**
>
> I have added the previous `ServicesService` class manually. Some TDD practitioners like to code-generate this file while they are writing their unit test rather than writing it first. Any method is fine as long as you are more productive. I chose to create the file first because sometimes VS creates this file in a different directory from where I intend it to.

I will create my unit tests class, which is called `ServicesServiceTests`, with the following code:

```
public class ServicesServiceTests : IDisposable
{
    private readonly ApplicationContextFakeBuilder _ctxBldr
        = new();
    private ServicesService? _sut;
```

```
public void Dispose()
{
    _ctxBldr.Dispose();
}
}
```

I have added `ApplicationContextFakeBuilder` immediately because I know that I am going to be dealing with sample data in my unit tests.

Now, I need to think of what I need from my service and build a unit test accordingly. The straightforward way to start is to pick the easiest scenario. If we have no barber service, then no service is returned:

```
[Fact]
public async Task
GetActiveServices_NoServiceInTheSystem_NoServices()
{
    // Arrange
    var ctx = _ctxBldr.Build();
    _sut = new ServicesService(ctx);

    // Act
    var actual = await _sut.GetActiveServices();

    // Assert
    Assert.True(!actual.Any());
}
```

I have decided in the test that there will be a method named `GetActiveServices`, and when this method is called, it will return a collection of active services. At this stage, the code doesn't compile; as such, a method doesn't exist. We have got our TDD's fail!

Now, we can instruct VS to generate this method, and then we can write the implementation:

```
public class ServicesService
{
    private readonly ApplicationContext _context;

    public ServicesService(ApplicationContext context)
    {
        _context = context;
```

```
        }

        public async Task<IEnumerable<Service>>
            GetActiveServices()
                => await _context.Services!.ToArrayAsync();
}
```

This is getting, through EF, all the available services, and since we did not store any service in the sample data, none will return.

If you run the test again, it will pass. This is our TDD's test pass. There is no need for the refactor stage, as this is a simple implementation. Congratulations, you have finished your first test!

> **Note**
> This test is simple, and it seems like a waste of time. However, this is a valid test case, and it also helps us create our domain class and inject the right dependencies. Starting with a simple test helps to progress in steady steps.

Adding the second unit test via TDD

The second feature that we need to add is the ability to get the active services only, not the ones that are no longer active, as they are no longer provided by the barber. So, let's start with this unit test:

```
[Fact]
public async Task
  GetActiveServices_TwoActiveOneInactiveService_TwoServices()
{
    // Arrange
    var ctx = _ctxBldr
        .WithSingleService(true)
        .WithSingleService(true)
        .WithSingleService(false)
        .Build();
    _sut = new ServicesService(ctx);
    var expected = 2;

    // Act
    var actual = await _sut.GetActiveServices();
```

```
      // Assert
      Assert.Equal(expected, actual.Count());
}
```

Our `Arrange` will add three services – two active and one inactive. It is interesting to see the code of `WithSingleService`:

```
public ApplicationContextFakeBuilder WithSingleService
      (bool isActive)
{
      _context.Add(new Service{ IsActive = isActive });
      return this;
}
```

If we run the test, of course, it will fail, as we have not added any filtration functionality to our service. Let's go ahead and add filtration to the service:

```
public async Task<IEnumerable<Service>> GetActiveServices()
      => await _context.Services!.Where(x => x.IsActive)
                                    .ToArrayAsync();
```

We have added a `Where` LINQ statement, which will do the trick. Run the tests again, and this test shall pass.

This was an easy requirement. In fact, all the stories are straightforward except story number 5. We will not list the other stories here because they are similar, but you can find them in the companion source code. Instead, we will focus on story number 5 as its complexity matches real-life production code and reveals the main benefit of TDD.

Implementing the fifth story (time management)

This story is about a time management system. It tries to manage barbers' time fairly, taking rest time into consideration. If you take a moment to think about this story, it is a complex one with many edge cases.

This story reveals the power of TDD as it will help you find a starting point and adds little incremental steps to build the requirement. When you finish, you will notice that you have automatically documented the story in the unit tests.

In the next sections, we will find a way to start from the easier-to-implement scenarios and climb up to more sophisticated test scenarios.

Checking for records

One gentle way to start our implementation that will make us think of the signature of the method is checking the parameters.

Logically, to determine an employee's availability, we need to know who this employee is by using `employeeId` and the length of time required. The length can be acquired from the service by `serviceId`. A logical name for the method can be `GetAvailableSlotsForEmployee`. Our first unit test is this:

```
[Fact]
public async Task
  GetAvailableSlotsForEmployee_ServiceIdNoFound_
    ArgumentException()
{
    // Arrange
    var ctx = _contextBuilder
        .Build();
    _sut = new SlotsService(ctx, _nowService, _settings);

    // Act
    var exception = await
        Assert.ThrowsAsync<ArgumentException>(
        () => _sut.GetAvailableSlotsForEmployee(-1));

    // Assert
    Assert.IsType<ArgumentException>(exception);
}
```

It doesn't compile; it is a fail. So, create the method in `SlotsService`:

```
public async Task<Slots> GetAvailableSlotsForEmployee(
    int serviceId)
{
    var service = await _context.Services!
        .SingleOrDefaultAsync(x => x.Id == serviceId);
    if (service is null)
    {
        throw new ArgumentException("Record not found",
```

```
            nameof(serviceId));
    }
    return null;
}
```

Now that you have the implementation in place, run the tests again, and they will pass. You can do the same for `employeeId` and do what we did for `serviceId`.

Starting with the simplest scenario

Let's add the simplest possible business logic to start with. Let's assume that the system has one employee called Tom. Tom has no shifts available in the system. Also, the system has one service only:

```
[Fact]
public async Task GetAvailableSlotsForEmployee_
    NoShiftsForTomAndNoAppointmentsInSystem_NoSlots()
{
    // Arrange
    var appointmentFrom =
        new DateTime(2022, 10, 3, 7, 0, 0);
    _nowService.Now.Returns(appointmentFrom);
    var ctx = _contextBuilder
        .WithSingleService(30)
        .WithSingleEmployeeTom()
        .Build();
    _sut = new SlotsService(ctx, _nowService, _settings);
    var tom = context.Employees!.Single();
    var mensCut30Min = context.Services!.Single();

    // Act
    var slots = await
        _sut.GetAvailableSlotsForEmployee(
        mensCut30Min.Id, tom.Id);

    // Assert
    var times = slots.DaysSlots.SelectMany(x => x.Times);
    Assert.Empty(times);
}
```

This will fail, as we have `null` returned by the method, whatever the input is. We need to continue adding bits of code to the solution. We can start with the following code:

```
...
var shifts = _context.Shifts!.Where(
    x => x.EmployeeId == employeeId);
if (!shifts.Any())
{
    return new Slots(Array.Empty<DaySlots>());
}
return null;
```

The previous code is exactly what is required to pass the test. The test is green now.

Elevating scenarios' complexity

The rest of the unit tests follow the same way of elevating test scenario complexity slightly. Here are other scenarios you might want to add:

```
[Theory]
[InlineData(5, 0)]
[InlineData(25, 0)]
[InlineData(30, 1, "2022-10-03 09:00:00")]
[InlineData(35, 2, "2022-10-03 09:00:00",
   "2022-10-03 09:05:00")]
public async Task GetAvailableSlotsForEmployee_
OneShiftAndNoExistingAppointments_VaryingSlots(
    int serviceDuration, int totalSlots,
        params string[] expectedTimes)
{
...
```

The previous test is, in fact, multiple tests (because we are using `Theory`) with each `InlineData` elevating complexity. As usual, do the red then green to let it pass before adding another suite of tests:

```
public async Task GetAvailableSlotsForEmployee_
   OneShiftWithVaryingAppointments_VaryingSlots(
      string appointmentStartStr, string appointmentEndStr,
      int totalSlots, params string[] expectedTimes)
```

```
{
    ...
```

This is also a test with multiple `InlineData`. Obviously, we cannot fit all the code here, so please have a look in `SlotsServiceTests.cs` for the complete unit tests.

As you start adding more test cases, whether by using `Theory` with `InlineData` or using `Fact`, you will notice that the code complexity in the implementation is going up. This is all right! But, do you feel the readability is suffering? Then it is time to refactor.

Now you have the advantage of unit tests protecting the code from being broken. Refactoring when the method is doing what you want it to do is part of the Red-Green-Refactor mantra. In fact, if you look at `SlotsService.cs`, I did refactor to improve readability by creating multiple private methods.

This story is complex, I will give you that. I could have picked an easier example, and everybody would be happy, but real-life code has ups and downs and varies in complexity, so I wanted to include one sophisticated scenario following the pragmatism theme of the book.

After this section, you might have some questions. I hope I am able to answer some of them below.

Answering frequently asked questions

Now that we have written the unit tests and the associated implementation, let me explain the process.

Are these unit tests enough?

The answer to this question depends on your target coverage and your confidence that all cases are covered. Sometimes, adding more unit tests increases the future maintenance overhead, so with experience, you would strike the right balance.

Why didn't we unit test the controllers?

The controllers should not contain business logic. We pushed all the logic to the services, then tested the services. What is left in the controllers is minimal code concerned with mapping different types to each other. Have a look at the controllers in `Uqs.AppointmentBooking.WebApi/ Controllers` to see what I mean.

Unit tests excel in testing business logic or areas where there are conditions and branching. The controllers in the coding style that we chose do not have that.

The controllers should be tested but through a different type of test.

Did we test the system enough?

No, we didn't! We did the unit tests part. We have not tested the controllers or the boot of the system (the content of `Program.cs`) and other small bits of the code.

We did not test them via unit tests as they are not business logic. However, they need testing, but unit tests are not the best testing type to check for the quality of these areas. You can cover these areas by other types of testing such as integration, Sintegration, and system tests.

We omitted testing some areas, how can we achieve high coverage?

Some areas of the code are not unit tested, such as `Program.cs` and the controllers. If you are aiming for high code coverage, such as 90%, you might not achieve it via unit testing alone, as there is a good amount of code that went here, in this chapter.

Achieving coverage by unit tests alone is unfair as you need additional testing types to achieve more coverage, or the developers would start cheating by adding meaningless tests to boost coverage. These tests do more harm than good as they will create a maintenance overhead.

Coverage calculation should include other types of tests, rather than relying on units alone. If this is the case, 90% is a realistic target and can lead to a high-quality product.

Sometimes it is hard to configure a coverage meter tool to measure the sum of multiple test types, so in this case, it makes sense to lower your coding coverage target to maybe 80% or so.

Summary

We have seen implementing realistic stories by setting up the system with EF and SQL Server, then building it a bit at a time by incrementally adding unit tests and increasing the complexity with every additional unit test.

We have seen a realistic fake test double and a concrete builder to construct our sample data.

We had to select multiple important scenarios to encourage you to examine the full source code, otherwise, the pages will be filled with code.

If you have read and understood the code, then I assure you that this is the peak of the complexity, as other chapters should be easier to read and follow. So congratulations, you have made it through the hard part of this book! I trust you can now go ahead and start your TDD-based project with EF and a relational DB.

Hopefully, this chapter has given you a guide to starting your new EF and SQL Server-based project. The next chapter does the same implementation but focuses on document DB and has different patterns than this one.

10

Building an App with Repositories and Document DB

In *Chapter 8*, *Designing an Appointment Booking App*, we laid the technical specifications and the design decisions for building an appointment booking system for a barber salon called Heads Up Barbers. This chapter is a continuation of *Chapter 8*, so I strongly advise you to be familiar with it first.

This chapter will implement the requirements in TDD style and will use the repository pattern with **Azure Cosmos DB**. The implementation will be applicable to other **document databases** aka **NoSQL** such as **MongoDB**, **Amazon DynamoDB**, **GCP Firestore**, and others.

If you are a fan of document DBs or you are using one at work, then this chapter is for you, while if you are using a relational database, then you might want to skip this chapter and go back to the previous chapter, *Chapter 9*. Both chapters, *Chapter 9* and *Chapter 10* have the same outcome, but they use different backend DB categories.

The chapter assumes you are familiar with a document DB service and the idea behind document DBs, not necessarily Cosmos DB, as from a TDD perspective, the implementation between different DB products is almost identical.

In this chapter, we will cover the following:

- Planning the code and the project structure
- Implementing the web APIs with TDD
- Answering frequently asked questions

By the end of the chapter, you will have experienced the implementation of an end-to-end app using TDD with mocks and document DB backend. Also, you will witness the analysis process that precedes writing unit tests.

Technical requirements

The code for this chapter can be found at the following GitHub repository:

`https://github.com/PacktPublishing/Pragmatic-Test-Driven-Development-in-C-Sharp-and-.NET/tree/main/ch10`

To run the project, you will need to have an instance of Cosmos DB installed. This can be one of the following:

- Azure Cosmos DB on the cloud under an Azure account
- **Azure Cosmos DB Emulator**, which can be installed locally on Windows, Linux, and macOS and can run from Docker

The implementation doesn't use any advanced Cosmos feature, so feel free to use any Cosmos flavor. I have tested the application with Azure Cosmos DB Emulator on Windows locally. You can find more about it here:

`https://docs.microsoft.com/en-us/azure/cosmos-db/local-emulator`

After installing the local emulator, you need to grab the connection string, which you can do by browsing to `https://localhost:8081/_explorer/index.html` and copying the connection string from the **Primary Connection String** field:

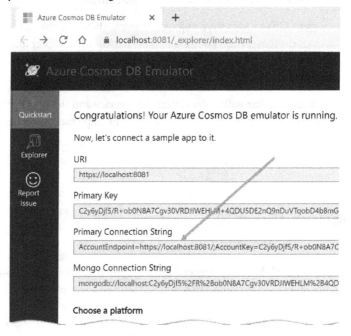

Figure 10.1 – Finding Cosmos DB connection string

To run the project, you have to set the connection string to your specific DB instance in `Uqs.AppointmentBooking.WebApi/AppSettings.json`, as follows:

```
"ConnectionStrings": {
    "AppointmentBooking": "[The primary connection string]"
},
```

The connection string is pointing to `LocalMachine` and will connect to a database called `AppointmentBooking`.

> **Note**
>
> I am not concerned in this chapter regarding setting multiple environments, but you may want to create multiple `AppSettings` for different environments and change the connection string accordingly.

The previous DB configuration steps are optional. You can implement the requirements for this chapter without using a DB, but you won't be able to run the project and interact with it in the browser.

Planning your code and project structure

In *Chapter 8, Designing an Appointment Booking App*, we planned our domain and analyzed what we needed to do. The project architecture will follow the classical three-tier applications of the client application (the website), business logic (the web APIs), and database (Cosmos DB). Let's translate this into VS solutions and projects.

In this section, we will create the solution and create the projects, and wire up the components.

Analyzing projects' structure

Get a group of senior developers to agree on one project structure, and you will end up with multiple structures! In this section, we will discuss a way of organizing your project structure that I have developed over the years.

Given that we are first going to build a website for the user and later a mobile app (not covered in this book), it makes sense to isolate the business logic to a web API project that can be shared by both the website and the mobile app. So, we will build a website project based on Blazor WebAssembly called `Uqs.AppointmentBooking.Website`.

The domain logic will be exposed as APIs, so we will create an ASP.NET API project for this one called `Uqs.AppointmentBooking.WebApi`.

The previous two projects needed to exchange data in an agreed structure called **data transportation objects (DTOs)**, commonly known as *contracts*. So, we will create a .NET library project called `Uqs.AppointmentBooking.Contracts`. This project will be referenced by both the website and the web API projects.

The web API project translates web requests into something we can understand in C#. In technical terms, this will manage our communication layer of HTTP with RESTful style APIs. So, the WebApi project will not contain business logic. The business logic will be in our domain project. We will create a domain project called `Uqs.AppointmentBooking.Domain`.

Your business logic will live in two places – the UI and the domain layer. The UI business logic will manage UI functionalities such as toggling dropdowns, blocking calendar days, responding to drag and drop, and disabling/enabling buttons, among others. This logic will live in the website project.

> **Important Note**
>
> UI frameworks such as Blazor and Angular act as standalone applications. These frameworks facilitate, by design, using a design pattern called **Model View View-Model** (**MVVM**), which makes dependency injection and, therefore, unit testing easy. However, unit testing the UI-specific elements (razor files in Blazor) requires a more specialized framework such as **bUnit**.

The language used in writing the code depends on the UI framework in use, such as Angular, React, and Blazor. In our implementation, we will have little code in the UI layer, so we will not be doing any UI unit tests.

The sophisticated business logic will live in the domain layer, and we will be writing it following the concepts of TDD. So, we shall create a project that will hold our domain unit tests and call it `Uqs.AppointmentBooking.Domain.Tests.Unit`.

To put these projects into perspective and map them to our three-tier architecture, we can have the following diagram:

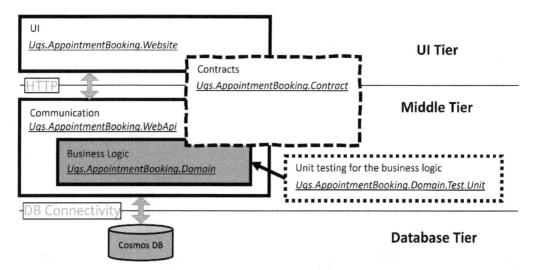

Figure 10.2 – The relationship between the projects and the application design

The previous diagram shows the functionality that each project provides to form the three-tier application. Let's start by creating the VS solution structure.

Creating projects and configuring dependencies

This is the boring inevitable part, creating the solution and the projects and then linking them together. In the following section, we will take the command line approach rather than the UI.

> **Note**
>
> I have added a text file, `create-projects.bat`, to the project source control that contains all the command lines, so you don't have to write them manually. You can copy and paste this file to your desired directory, then, from your command line, execute the file.

The following is the list of commands that will create your VS solution and its projects:

1. From your OS console, navigate to the directory where you want to create your new solution and execute the following to create the solution file:

```
md UqsAppointmentBooking
cd UqsAppointmentBooking
dotnet new sln
```

2. Execute this to create the projects and notice that we are using a different template for each project:

```
dotnet new blazorwasm -n
  Uqs.AppointmentBooking.Website
dotnet new webapi -n Uqs.AppointmentBooking.WebApi
dotnet new classlib -n Uqs.AppointmentBooking.Contract
dotnet new classlib -n Uqs.AppointmentBooking.Domain
dotnet new xunit -n
  Uqs.AppointmentBooking.Domain.Tests.Unit
```

3. Add the projects to the solution:

```
dotnet sln add Uqs.AppointmentBooking.Website
dotnet sln add Uqs.AppointmentBooking.WebApi
dotnet sln add Uqs.AppointmentBooking.Contract
dotnet sln add Uqs.AppointmentBooking.Domain
dotnet sln add Uqs.AppointmentBooking.Domain
  .Tests.Unit
```

4. Now let's set up dependencies between projects:

```
dotnet add Uqs.AppointmentBooking.Website reference
   Uqs.AppointmentBooking.Contract
dotnet add Uqs.AppointmentBooking.WebApi reference
   Uqs.AppointmentBooking.Contract
dotnet add Uqs.AppointmentBooking.Domain reference
   Uqs.AppointmentBooking.Contract
dotnet add Uqs.AppointmentBooking.WebApi reference
   Uqs.AppointmentBooking.Domain
dotnet add Uqs.AppointmentBooking.Domain.Tests.Unit
   reference Uqs.AppointmentBooking.Domain
```

And the last bit is adding the required NuGet packages to the project. The domain project will communicate with Cosmos DB using the Cosmos SDK from the `Microsoft.Azure.Cosmos` package. Add this library to the `Domain` project as such:

```
dotnet add Uqs.AppointmentBooking.Domain package
   Microsoft.Azure.Cosmos
```

5. The unit testing project will require `NSubstitute` for mocking, so let's add its NuGet:

```
dotnet add Uqs.AppointmentBooking.Domain.Tests.Unit
   package NSubstitute
```

For visual inspection, you can open the solution file with VS, and it should look as such:

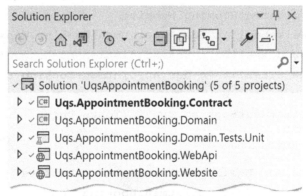

Figure 10.3 – A view of VS Solution Explorer

At this stage, your solution structure should look similar.

Now that the project structure is in place, we will modify the code.

Setting up the domain project

From the domain analysis in *Chapter 8, Designing an Appointment Booking App*, we have created a list of the domain objects. I will not go through them again; I will just create and add them to the Domain project under DomainObjects:

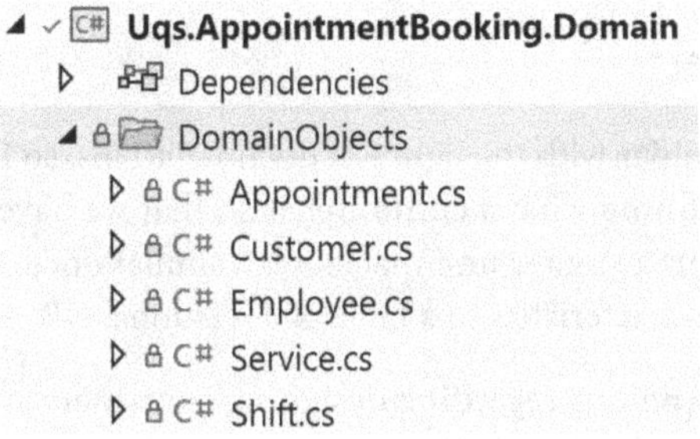

Figure 10.4 – Added domain objects

These are just data structures with no business logic. Here is the source code of one of them, the Customer domain object:

```
namespace Uqs.AppointmentBooking.Domain.DomainObjects;
public class Customer : IEntity
{
    public string? Id { get; set; }
    public string? FirstName { get; set; }
    public string? LastName { get; set; }
}
```

You can view the rest of the files in the chapter's GitHub repository online. This is the interface of the previous class:

```
public interface IEntity
{
    public string? Id { get; set; }
}
```

IEntity is an interface that ensures every domain object that is going to be persisted to a document container has an Id.

> **Note**
>
> Id is a string as this is what a document DB is expecting and usually, but not necessarily, the string is a GUID.

What is the relationship between our document containers and our domain objects?

Designing your containers

I am assuming that you are familiar with the basics of document DBs, so I won't be going into much detail. Let me start first by defining what a **container** is so that we have the same understanding across the chapter. A container is a storage unit that stores a similar document type. A container in a document DB shares similar characteristics to a table of a relational DB.

There are many schools and opinions regarding designing containers and the factors to take into consideration, but our focus in this book is TDD, so we shall keep this short and to the point. Obviously, designing in DDD, while it has some guidelines, is still a subjective process. It feels that our aggregate routes, Service, Employee, Customer, and Appointment, are direct contenders for becoming containers, so we will set them to be.

Next, we need a way for our domain services to interact with the database. This is done through the repository pattern.

Exploring the repository pattern

Our containers are now defined. We just need the mechanism of interacting with these containers. DDD employs the repository pattern for this purpose. Let's shed some light on the role of the pattern and where it sits in our application.

Understanding the repository pattern

The repository layer is code that knows how to interact with a database, it doesn't matter what type of database is underlying (it doesn't matter whether this is Cosmos, SQL Server, a text file, or others), document, relational, or others. The layer is meant to isolate the domain layer from understanding the specifics of the database. Instead, the domain services will only be concerned about what data to persist rather than how they will be persisted.

The next diagram shows the repository as the bottom layer in the domain layer:

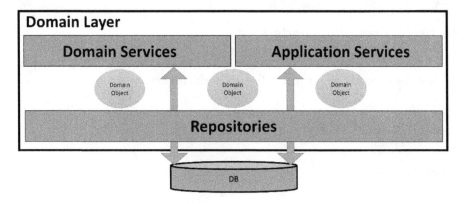

Figure 10.5 – The repositories within DDD

You can see from the diagram that any domain object to be persisted in the database passes through the repositories layer.

The common practice of creating a repository is creating a repository per container. So, for our application, we will have four repositories:

- `ServiceRepository`
- `CustomerRepository`
- `AppointmentRepository`
- `EmployeeRepository`

Given that we will have to unit test our implementation, our repositories need to be unit test-ready.

Repositories and unit testing

We suddenly started speaking about repositories in a chapter about TDD. The reason is that when you think of unit testing, the first thing that comes to mind is dependencies and how to isolate the database.

Repositories are the answer to this question, as they should provide the abstraction necessary to convert the database into an injectable dependency. You will see this clearly later on in this chapter.

> **Note**
>
> If you have worked with a relational database using an **object-relational mapper (ORM)** such as **Entity Framework** or **NHibernate**, then you have probably not used the repository pattern directly as the ORM frameworks eliminate the need to use it.

You will see that our repositories will have interfaces that will allow them to be injection-ready. Enough of the theory and let me show you some code.

Implementing the repository pattern

Now that you have an idea about what repositories are, let's start with an example. One of the repositories needed is `ServiceRepository`, which will interact with the Service repository:

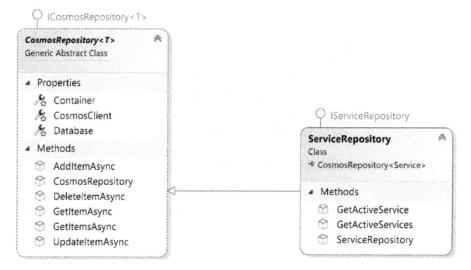

Figure 10.6 – The Service repository

The `ServiceRepository` class contains methods for adding a service, deleting it and searching for a specific service, and more. Let's pick a random method from the `GetActiveService` repository class:

```
public async Task<Service?> GetActiveService(string id)
{
  var queryDefinition = new QueryDefinition(
  "SELECT * FROM c WHERE c.id = @id AND c.isActive = true")
          .WithParameter("@id", id);
  return (await GetItemsAsync(queryDefinition))
   .SingleOrDefault();
}
```

The method above uses Cosmos DB-specific code to access the database and returns a service by its ID.

Note that the repository is implementing the `IServiceRepository` interface, which will become handy later during unit testing.

There is plenty of repetition in how a repository interacts with a container. It stores a document, it reads a document, it deletes a document, it searches for a document, and so forth. So, we can create a small framework to embed these behaviors and reduce the repeated code.

Utilizing a repository pattern framework

Every time I've seen a project accessing a document DB, I notice that the developers create a small repository framework in advance to simplify the code. Here is an excerpt from a framework that I created to access Cosmos DB, the `CosmosRepository<T>` class, which is inherited by all repositories:

```csharp
using Microsoft.Azure.Cosmos;
using Microsoft.Extensions.Options;
using System.Net;
namespace Uqs.AppointmentBooking.Domain.Repository;
public abstract class CosmosRepository<T> :
    ICosmosRepository<T> where T : IEntity
{

    protected CosmosClient CosmosClient { get; }
    protected Database Database { get; }
    protected Container Container { get; }

    public CosmosRepository(string containerId,
    CosmosClient cosmosClient,
    IOptions<ApplicationSettings> settings)
    {
        CosmosClient = cosmosClient;
        Database = cosmosClient.GetDatabase(
            settings.Value.DatabaseId);
        Container = Database.GetContainer(containerId);
    }

    public Task AddItemAsync(T item)
    {
        return Container.CreateItemAsync(item,
            new PartitionKey(item.Id));
```

```
    }
...
```

The code above provides the repository with the basic methods required to interact with the database, such as `AddItemAsync`.

Going into details about the specifics of Cosmos DB is beyond the scope of the book, but the code is easy to read and you can find the complete implementation in the `Uqs.AppointmentBooking.Domain/Repository` directory in the source code.

Now that we have created the repositories, for development purposes, we might need some test data to fill the pages with some meaningful data. We shall do this next.

Adding seed data

Newly created DBs have empty containers, and the `seed` class is meant to pre-populate the tables with sample data.

I will not list the code here as it is outside the scope of the chapter, but you can look at the code in the `Domain` project in `Database/SeedData.cs`.

We've just finished the setup for the `WebApi` project that is going to be consumed by the website, so let's create the website next.

Setting up the website project

Phase I of this implementation includes creating a website to access the APIs to provide a UI for the user, which we did previously in this chapter by command line. However, website implementation is outside the scope of this chapter and the book, in general, as it is not related to TDD, so I will not be going through the code.

Although, we are interested in one aspect – what does `Website` require from `WebApis`? We will need to understand this in order to build the required functionality in `WebApis` the TDD way.

We will answer this question bit by bit in the next section of this chapter.

In this section, we did the setup and configuration aspect of the project, and we have not done anything that is affected by TDD. You may have noticed that I referred you to the companion source code on multiple occasions, as I wanted to keep the focus on the next section while still providing you with the source code.

Implementing the WebApis with TDD

To build the WebApi project, we are going to look at each requirement from *Chapter 8, Designing an Appointment Booking App*, and provide the implementation that satisfies it using the TDD style.

The requirements are all stated in terms of `Website` and its functionality, and they do not dictate how to build our APIs. `Website` will have to call `WebApis` for any business logic as it has no access to the DB and does UI-related business logic only.

In this section, we will go through working in TDD mode, taking into consideration our persistence provider, the repositories.

Implementing the first story

The first story in our requirement is very easy. The website is going to display all the available services that we have. Since the website will request this data from `WebApi` through a RESTful API call, the domain layer will have a service that will return this list. Let's dig further if the website is to display this:

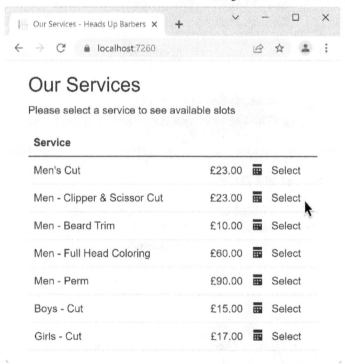

Figure 10.7 – A UI of the requirements of Story 1

It will need to issue a RESTful call to the WebApi, which can look as follows:

```
GET https://webapidomain/services
```

This UI will require a few data properties that should be returned by this API. So, the fetched JSON can look like an array of this:

```json
{
    "ServiceId": "e4c9d508-89d7-49cd-86c2-835cde94472a",
    "Name": "Men - Clipper & Scissor Cut",
    "Duration": 30,
    "Price": 23.0
}
```

You can see where each part is used on the page, but maybe `ServiceId` is not very clear. It will be used to construct the URL of the *Select* hyperlink. So, we can now design the contract type that will render this JSON, which could look like this:

```
namespace Uqs.AppointmentBooking.Contract;
public record Service(string ServiceId, string Name,
    int Duration, float Price);
```

This `record` contract will render the previous JSON code, and the full returned array contract could look like this:

```
namespace Uqs.AppointmentBooking.Contract;
    public record AvailableServices(Service[] Services);
```

You can find these contract types and all the other contracts in the `Contract` project.

Adding the first unit test via TDD

Thinking along the lines of DDD, we will have a domain service called `ServicesService`, which will handle retrieving all the available services. So, let's have the structure of this service. We will create it in the `Domain` project under `Services`. Here is the code:

```
public class ServicesService
{
}
```

There is nothing special here. I have just helped VS understand when I type `ServicesService`, it will guide me to this class.

> **Note**
>
> I have added the previous `ServicesService` class manually. Some TDD practitioners like to code-generate this file while they are writing their unit test rather than writing it first. Any method is fine as long as you are more productive. I chose to create the file first because sometimes VS creates this file in a different directory from where I intend to.

I will create my unit tests class, which is called `ServicesServiceTests`, with the following code:

```
public class ServicesServiceTests
{
    private readonly IServiceRepository _serviceRepository
        = Substitute.For<IServiceRepository>();
    private ServicesService? _sut;
}
```

I have added `IServiceRepository` immediately because I know that I am going to be dealing with the database in my unit tests and this interface is going to be my mocked dependency.

Now, I need to think of what I need from my service and build a unit test accordingly. The straightforward way to start is to pick the easiest scenario. If we have no barber service, then no service is returned:

```
[Fact]
public async Task
  GetActiveServices_NoServiceInTheSystem_NoServices()
{
    // Arrange
    _sut = new ServicesService(_serviceRepository);

    // Act
    var actual = await _sut.GetActiveServices();

    // Assert
    Assert.True(!actual.Any());
}
```

I have decided in the test that there will be a method named `GetActiveServices`, and when this method is called, it will return a collection of active services. At this stage, the code doesn't compile; as such, a method doesn't exist. We have got our TDDs fail!

Now, we can instruct VS to generate this method, and then we can write the implementation:

```
public class ServicesService
{
    private readonly IServiceRepository _serviceRepository;

    public ServicesService(
        IServiceRepository serviceRepository)
    {
        _serviceRepository = serviceRepository;
    }

    public async Task<IEnumerable<Service>>
        GetActiveServices() =>
        await _serviceRepository.GetActiveServices();
}
```

This is getting, through the repository, all the available services, and since the repository is not mocked to return any service, an empty collection will return.

If you run the test again, it will pass. This is our TDD test pass. There is no need for the refactor stage, as this is a simple implementation. Congratulations, you have finished your first test!

> **Note**
>
> This test is simple, and it seems like a waste of time. However, this test is a valid test case, and it also helps us create our domain class and inject the right dependencies. Starting with a simple test helps to progress in steady steps.

Adding the second unit test via TDD

The second feature that we need to add is the ability to get active services. So, let's start with this unit test:

```
[Fact]
public async Task
  GetActiveServices_TwoActiveServices_TwoServices()
{
    // Arrange
    _serviceRepository.GetActiveServices()
        .Returns(new Service[] {
```

```
            new Service{IsActive = true},
            new Service{IsActive = true},
        });
    _sut = new ServicesService(_serviceRepository);
    var expected = 2;

    // Act
    var actual = await _sut.GetActiveServices();

    // Assert
    Assert.Equal(expected, actual.Count());
}
```

What is interesting here is the way we are mocking the GetActiveServices repository method. The method is mocked to return an array of Service when the service calls it. This is how we have substituted the database for the relevant repository.

If you run this, it should pass from the first time without failure, so it is not going to fail then pass. It just happened to be this way. In this scenario, I would debug my code to see why the unit test passed without me implementing the code, and it is obvious the implementation code for the first unit test made was enough to cover the second scenario.

This was an easy requirement. In fact, all the stories are straightforward except story number 5. We will not list the other stories here because they are similar, but you can find them in the companion source code. Instead, we will focus on story number 5 as its complexity matches a real-life production code and would reveal the main benefit of TDD.

Implementing the fifth story (time management)

This story is about a time management system. It tries to manage the barbers' time fairly, taking rest time into consideration. If you take a moment to think about this story, it is a complex one with many edge cases.

This story reveals the power of TDD as it will help you find a starting point and adds little incremental steps to build the requirement. When you finish, you will notice that you have automatically documented the story in the unit tests.

In the next sections, we will find a way to start with the easier-to-implement scenarios and climb up to more sophisticated test scenarios.

Checking for records

One gentle way to start our implementation that will make us think of the signature of the method is by checking the parameters.

Logically, to determine an employee's availability, we need to know who this employee is by using `employeeId` and the length of the time required. The length can be acquired from the service by `serviceId`. A logical name for the method can be `GetAvailableSlotsForEmployee`. Our first unit test is this:

```
[Fact]
public async Task
  GetAvailableSlotsForEmployee_ServiceIdNoFound_
    ArgumentException()
{
    // Arrange

    // Act
    var exception = await
        Assert.ThrowsAsync<ArgumentException>(() =>
        _sut.GetAvailableSlotsForEmployee("AServiceId"));

    // Assert
    Assert.IsType<ArgumentException>(exception);
}
```

It doesn't compile; it is a fail. Now create the method in `SlotsService`:

```
public async Task<Slots> GetAvailableSlotsForEmployee(
    string serviceId)
{
    var service = await
        _serviceRepository.GetItemAsync(serviceId);
    if (service is null)
    {
        throw new ArgumentException("Record not found",
        nameof(serviceId));
    }
```

```
        return null;
    }
}
```

Now that you have the implementation in place, run the tests again, and they will pass. You can do the same for `employeeId` and follow what we did for `serviceId`.

Starting with the simplest scenario

Let's add the simplest possible business logic to start with. Let's assume that the system has one employee called Tom. Tom has no shifts available in the system. Also, the system has one service only:

```
[Fact]
public async Task GetAvailableSlotsForEmployee_
    NoShiftsForTomAndNoAppointmentsInSystem_NoSlots()
{
    // Arrange
    var appointmentFrom = new DateTime(
        2022, 10, 3, 7, 0, 0);
    _nowService.Now.Returns(appointmentFrom);
    var tom = new Employee { Id = "Tom", Name =
        "Thomas Fringe", Shifts = Array.Empty<Shift>() };
    var mensCut30Min = new Service { Id = "MensCut30Min",
        AppointmentTimeSpanInMin = 30 };
    _serviceRepository.GetItemAsync(Arg.Any<string>())
        .Returns(Task.FromResult((Service?)mensCut30Min));
    _employeeRepository.GetItemAsync(Arg.Any<string>())
        .Returns(Task.FromResult((Employee?)tom));

    // Act
    var slots = await
        _sut.GetAvailableSlotsForEmployee(mensCut30Min.Id,
        tom.Id);

    // Assert
    var times = slots.DaysSlots.SelectMany(x => x.Times);
    Assert.Empty(times);
}
```

You can see how the repositories are being populated via mocking. This is how we set up our database and do the dependency injection. We were able to do this as `SlotsService` is accessing the database through the repositories and if the repositories are mocked, then we have replaced our database.

> **Note**
>
> Replacing the database with mocked repositories is a hot interview question that goes like *how do you clean the database after each unit test?*. This is a trick question, as you don't interact with the database during the unit test and you mock your repositories instead. The question comes in multiple variations.

This will fail, as we have `null` returned by the method, whatever the input is. We need to continue adding bits of code to the solution. We can start with the following code:

```
...
if (!employee.Shifts.Any())
{
    return new Slots(Array.Empty<DaySlots>());
}
return null;
```

The previous code is exactly what is required to pass the test. The test is green now.

Elevating the scenarios' complexity

The rest of the unit tests follow the same way of elevating test scenario complexity slightly. Here are other scenarios you might want to add:

```
[Theory]
[InlineData(5, 0)]
[InlineData(25, 0)]
[InlineData(30, 1, "2022-10-03 09:00:00")]
[InlineData(35, 2, "2022-10-03 09:00:00",
    "2022-10-03 09:05:00")]
public async Task GetAvailableSlotsForEmployee_
  OneShiftAndNoExistingAppointments_VaryingSlots(
      int serviceDuration, int totalSlots,
        params string[] expectedTimes)
{
    ...
```

The previous test is, in fact, multiple tests (because we are using `Theory`) with each `InlineData` elevating complexity. As usual, do the red then green to let it pass before adding another suite of tests:

```
public async Task GetAvailableSlotsForEmployee_
  OneShiftWithVaryingAppointments_VaryingSlots(
    string appointmentStartStr, string appointmentEndStr,
      int totalSlots, params string[] expectedTimes)
{
  ...
```

This is also a test with multiple `InlineData`. Obviously, we cannot fit all the code here, so please have a look in `SlotsServiceTests.cs` for the complete unit tests.

As you start adding more test cases, whether by using `Theory` with `InlineData` or using `Fact`, you will notice that the code complexity in the implementation is going up. This is all right! Do you feel the readability is suffering? Then it is time to refactor.

Now you have the advantage of unit tests protecting the code from being broken. Refactoring when the method is doing what you want it to do is part of the *Red-Green-Refactor* mantra. In fact, if you look at `SlotsService.cs`, I did refactor to improve readability by creating multiple private methods.

This story is complex, I will give you that. I could have picked an easier example, and everybody would be happy, but real-life code has ups and downs and varies in complexity, so I wanted to include one sophisticated scenario following the pragmatism theme of the book.

After this section, you might have some questions. I hope I am able to answer some of them below.

Answering frequently asked questions

Now that we have written the unit tests and the associated implementation, let me explain the process.

Are these unit tests enough?

The answer to this question depends on your target coverage and your confidence that all cases are covered. Sometimes, adding more unit tests increases the future maintenance overhead, so with experience, you would strike the right balance.

Why didn't we unit test the controllers?

The controllers should not contain business logic. We pushed all the logic to the services, then tested the services. What is left in the controllers is minimal code concerned with mapping different types to each other. Have a look at the controllers in `Uqs.AppointmentBooking.WebApi/Controllers` to see what I mean.

Unit tests excel in testing business logic or areas where there are conditions and branching. The controllers in the coding style that we chose do not have that.

The controllers should be tested but through a different type of test.

Why didn't we unit test the repositories implementation?

The repositories contain specific code for Cosmos DB with minimal to no business logic. The code there is interacting with the SDK directly and testing it doesn't prove anything, as you will be making assumptions (through test doubles) about the behavior of the framework.

Sometimes a repository contains a bit of business logic, such as `ServiceRepository` picking the active services only, rather than all services. It is still hard to test this logic as it is embedded in SQL-like syntax, which is hard to unit test.

On the contrary, testing your repositories expands your unit tests footprint in a negative way, which makes your code more brittle.

Some developers still unit test their repositories for code coverage purposes, but the mistake here is that code coverage is the combination of all types of tests and not just unit tests. Your repositories should be covered by a different type of test such as Sintegration testing.

Did we test the system enough?

No, we didn't! We did the unit tests part. We have not tested the controllers or the boot of the system (the content of `Program.cs`) and other small bits of the code.

We did not test them via unit tests, as they are not business logic. However, they need testing, but unit tests are not the best testing type to check for the quality of these areas. You can cover these by other types of testing such as integration, Sintegration, and system tests, as per our discussion in *Chapter 4, Real Unit Testing with Test Doubles*.

We omitted testing some areas, how can we achieve high coverage?

Some areas of the code are not unit tested, such as `Program.cs` and the controllers. If you are aiming for high code coverage, such as 90%, you might not achieve it via unit testing alone, as there is a good amount of code that went here.

Achieving coverage by unit tests alone is unfair, or the developers would start cheating by adding meaningless tests to boost coverage. These tests do more harm than good as they will create a maintenance overhead.

Coverage calculation should include other types of tests, rather than relying on units alone. If this is the case, 90% is a realistic target and can lead to a high-quality product.

Sometimes it is hard to configure a coverage meter tool to measure the sum of multiple test types, so in this case, it makes sense to lower your coding coverage target to maybe 80% or so. Because not all tests run locally, a local test coverage tool (such as *Fine Code Coverage*, which was discussed earlier, in *Chapter 6, The FIRSTHAND Guidelines of TDD*), can only calculate the coverage for the locally executed tests.

So the short answer is your coverage should be made to include all your testing types, which takes some effort. Or you can reduce your coverage to unit test only and go for lower coverage.

Summary

We have seen implementing realistic stories by setting up the system with repositories and Cosmos DB, then building it a bit at a time by incrementally adding unit tests and increasing the complexity with every additional unit test.

We had to select multiple important scenarios to encourage you to examine the full source code. Otherwise, the pages will be filled with code.

If you have read and understood the code, then I assure you that this is the peak of the complexity, as other chapters should be easier to read and follow. So congratulations, you have made it through the hard part of this book! I trust you can now go ahead and start your TDD-based project with a document DB.

This chapter ended with the implementation of a realistic project based on TDD. Hopefully, by understanding this part of the book, you are equipped to write your TDD-based project with a relational DB or a document DB.

The next part of the book goes about introducing unit tests to your project and organization, dealing with existing legacy code, and building a continuous integration system. I call that the fun part, where you take your TDD knowledge and expand it.

Part 3:
Applying TDD to Your Projects

Now that we know how to build an application with TDD, we want to take the next step. In this part, we will cover how to combine unit testing with continuous integration, how to deal with legacy projects, and how to implement TDD in your organization. The following chapters are included in this part:

- *Chapter 11, Implementing Continuous Integration with GitHub Actions*
- *Chapter 12, Dealing with Brownfield Projects*
- *Chapter 13, The Intricacies of Rolling Out TDD*

11

Implementing Continuous Integration with GitHub Actions

You wrote unit tests and other types of tests and you are happy about your code coverage and quality. So far so good, but who is going to make sure these tests are going to run every time the code changes? Is it the developer that is pushing new code? What if they forget? What if there are merge problems in source control that might break your tests? Who is going to check?

You've already figured out the answer. It is the **continuous integration** (**CI**) system that you should have in place. CI is the natural companion to unit testing, and you can rarely find a modern project today without a CI system in place.

In this chapter, we will cover the following:

- An introduction to continuous integration
- Implementing a CI process with GitHub Actions

By the end of the chapter, you will be able to implement an end-to-end CI process with GitHub Actions.

Technical requirements

The code for this chapter can be found at the following GitHub repository:

https://github.com/PacktPublishing/Pragmatic-Test-Driven-Development-in-C-Sharp-and-.NET/tree/main/.github/workflows

Introduction to continuous integration

The idea behind the term **CI** is that new code is continuously integrated with existing code, which results in a system that can be shipped to production at any time (or at least this is the intention).

The route from software development to production is referred to as the **release pipeline**, where the code passes through multiple processes to reach production, such as compiling the code, deploying the binaries on the dev environment, allowing a QA to pull the code to a certain environment, and others. CI is an integral part of the release pipeline.

A CI system requires a host so that it can do various operations on the code. The host is a combination of a server and an operating system:

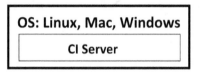

Figure 11.1 – CI server in an OS

Here are a few examples of on-premise CI servers:

- Cruise Control
- Team City
- **Team Foundation Server (TFS)**
- Jenkins
- Octopus Deploy

You will be able to find a SaaS solution for the previously mentioned system as well. However, today, the native cloud solutions are more popular, such as the following:

- GitHub Actions
- Azure DevOps
- AWS CodePipeline
- Octopus Cloud
- GitLab CI/CD

The concepts of these systems are the same and when you learn one, you can easily learn another. Now, let's see how a CI system works.

CI workflow

A CI system applies a workflow to your code by executing a series of actions. The actions are configurable and may vary between one project and another based on the project's needs. This is a generic workflow for a CI system:

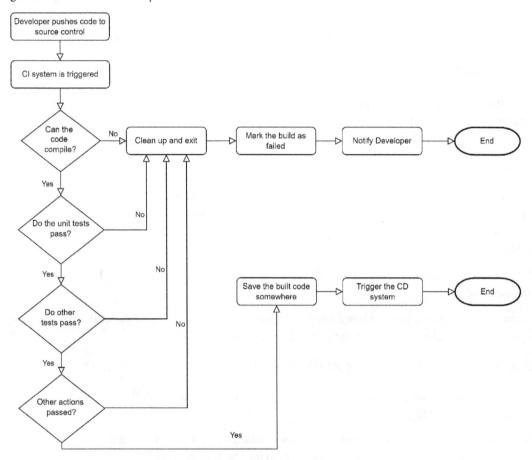

Figure 11.2 – CI workflow

When a developer pushes new code into the source control, the CI system is usually listening for this event. It takes a copy of the code and tries to compile it similarly to what happens when you compile on your machine. Then, it tries to discover all the unit tests in your test projects and execute them.

CI systems are highly customizable, so a developer might add execution for multiple types of tests and other housekeeping steps.

If any step fails, the CI system will abandon the build and mark it as failed, and notify the developer via a pre-configured way, such as an email.

If all the steps pass, then a **continuous deployment (CD)** system can be plugged into the end of this process. A CD system deploys your built code (binaries) into servers based on your specific preferences. Therefore, you always hear the terms *CI/CD* used simultaneously.

Next, we are going to discuss the importance of CI in a software engineering process.

The benefits of CI systems

CI systems are usually plugged in from sprint zero of an agile process. They exist from day one for good reasons. Obviously, building a CI system pipeline takes time and effort, so there should be a justification for the added effort. Here, we will see the importance and benefits of a CI system.

Having the code compiling at all times

How many times in a team project have you pulled the latest code from source control and discovered that it does not compile? The CI system ensures the code compiles at all times and the best practice is not to pull the new code if the CI flags a previous push as a failure, because you won't be able to continue working unless you fix the code yourself or wait for your colleague to fix it.

Of course, you will have to pull the broken build if you are intending to fix it, but the CI gives you an earlier indication.

Usually, broken code in source control happens when a developer doesn't pull the latest from source control, and compiles and executes the tests before pushing their code.

Having the unit tests passing at all times

Developers might forget to execute the unit tests on their machine before pushing their code, but the CI system won't forget!

As per the previous discussion in this book, where unit tests should be high-performance, the CI should take no time in running all your unit tests to feed back to the team that the build is safe and ready for them to pull the new code.

I have specifically mentioned unit tests as they provide fast feedback. You might have other types of tests that might take time to execute, and you may decide whether you want to execute them with every push or at certain times of the day. Other tests are usually slow and take minutes to execute, and you might want to run them in parallel but not block the feedback until they finish, which might be 10 minutes or an hour.

Compiling the code in a ready state for CD

If the build has compiled successfully and passed testing, then this is ready for manual testing if you have this as part of your software engineering process or if you have binaries ready to be deployed to your environments by the CD process.

The CD process takes the output of your CI and deploys it to configured locations such as your dev environment, UAT, and production.

CI is not optional anymore in today's software engineering process, as you can see from the benefits. There is no excuse not to implement one. It is cheap, and it is easy to implement, as we are going to see next when we utilize GitHub Actions.

Implementing a CI process with GitHub Actions

Initially, when I was designing the guidelines for the chapters in this book, I was planning to give a sample implementation using Azure DevOps, as it has a popular CI system. However, GitHub Actions climbed up fast and quickly became the developers' choice in configuring a CI system, so I changed my mind and I am going to use GitHub Actions instead.

GitHub Actions can deal with multiple programming stacks; one of them is .NET Core, which is what we are concerned about in this chapter.

Obviously, you will need to have a GitHub account for using GitHub Actions and you will be glad to know that the free tier gives you 2,000 minutes/month of running time, which should be enough for a small solo project.

Next, we are going to use GitHub Actions as a CI system for the project in *Chapter 10, Building an App with Repositories and Document DB* you don't need to have read the chapter, we just want a solution that has a project and unit tests against it so we can demonstrate how Actions works and *Chapter 10* has that.

Creating a sample project in a GitHub repo

To follow along, you will need to have a GitHub account with a GitHub repository hosting a .NET project. If you don't have one, then you can create a free GitHub account.

You will need the code of *Chapter 10* in your repository, so the fastest way to do that is to go on this book's GitHub page and hit **Fork | Create a new fork**:

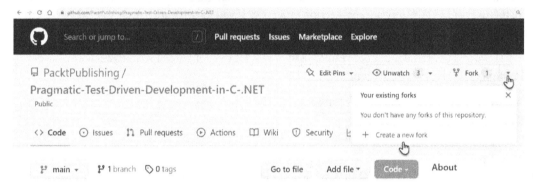

Figure 11.3 – Create a new fork

Or go to the repository URL and add `/fork` to the end, like this: `https://github.com/PacktPublishing/Pragmatic-Test-Driven-Development-in-C-Sharp-and-.NET/fork`. Then hit **Create fork**:

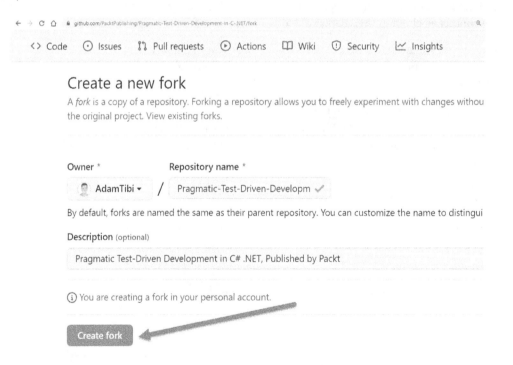

Figure 11.4 – Filling in the form and hitting Create fork

The fork will copy the content of a repository into a newly created repository that you own so you can play with the code without affecting the original repository.

While we've copied all the code, we are only interested in this earlier chapter (*Chapter 10*) in the solution, located in the following:

```
/ch10/UqsAppointmentBooking/UqsAppointmentBooking.sln
```

Now that you have the same code, we can create the GitHub Actions CI for this project.

Creating a workflow

First, to write any configuration for GitHub Actions, you will have to be familiar with YAML. YAML is a file format alternative to JSON that is geared for readability by humans. You will see examples of YAML as we go along.

Let's create a workflow for the *Chapter 10* project using the GitHub Actions wizard. From your GitHub repository, select **Actions** | **New workflow**:

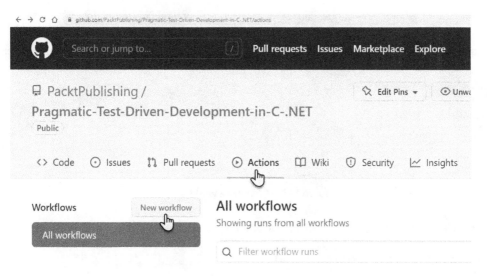

Figure 11.5 – Creating a new workflow

GitHub Actions comes back with a list of suggestions based on your repository content:

Get started with GitHub Actions

Build, test, and deploy your code. Make code reviews, branch management, and issue triaging work the way you want. Select a workflow to get started.

Skip this and set up a workflow yourself →

Q Search workflows

Suggested for this repository

.NET	Jekyll	.NET Desktop
By GitHub Actions	By GitHub Actions	By GitHub Actions
Build and test a .NET or ASP.NET Core project.	Package a Jekyll site using the jekyll/builder Docker image.	Build, test, sign and publish a desktop application built on .NET.
Configure C# ●	Configure HTML ●	Configure C# ●

Figure 11.6 – Suggestion list for a GitHub Actions workflow template

Our code is .NET Core, so the first suggestion is suitable; let's hit **Configure**. We will get this page:

Figure 11.7 – Creating a workflow

Note that GitHub has already suggested a file location for your workflow configuration:

/.github/workflows/dotnet.yml

GitHub Actions lives in the `workflows` directory. It also suggested the following YAML code for a start:

```
name: .NET

on:
  push:
    branches: [ "main" ]
```

```
  pull_request:
    branches: [ "main" ]

jobs:
  build:

    runs-on: ubuntu-latest

    steps:
    - uses: actions/checkout@v3
    - name: Setup .NET
      uses: actions/setup-dotnet@v2
      with:
        dotnet-version: 6.0.x
    - name: Restore dependencies
      run: dotnet restore
    - name: Build
      run: dotnet build --no-restore
    - name: Test
      run: dotnet test --no-build --verbosity normal
```

This workflow is called .NET and will be triggered when someone pushes code into the main branch or raises a pull request to the main branch.

The CI system uses the latest version of Ubuntu Linux available at GitHub Actions, which at the time of writing is Ubuntu 20.04. The OS will be hosting the build while various actions are applied to it. Linux is usually chosen by default because it is more efficient to run and cheaper than Windows, and obviously supports .NET Core.

Then the execution of the steps starts with the following:

1. actions/checkout@v3: This action checks out your repository, so your workflow can access it. This is requesting version 3 of this action.

2. actions/setup-dotnet@v2: Fetches .NET SDK using version 2 of this library and specifies .NET Core 6 as the .NET version. This allows us to use .NET CLI after.

3. dotnet restore: This is a standard .NET **command-line interface** (**CLI**) command to restore the NuGet packages.

4. dotnet build: Compiles the solution.

5. dotnet test: Executes all the test projects in the solution.

Steps 1 and *2* prepare your workspace on the host OS to be able to execute the `.NET CLI` command in the same way you would execute it on your local machine. As you've already figured out, the whole text uses YAML syntax.

You can go ahead and push the **Start commit** button:

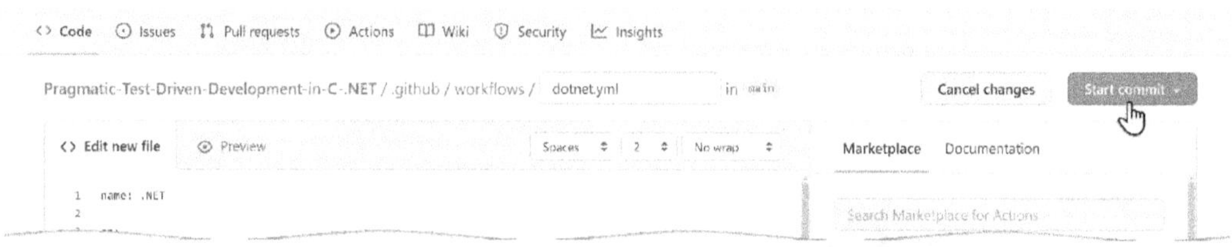

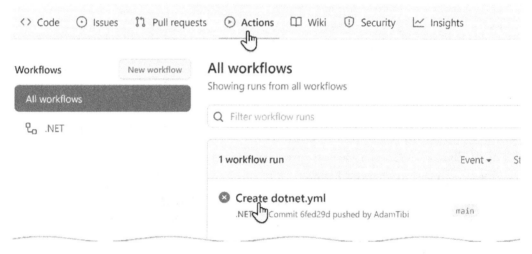

Figure 11.8 – Start commit

After this, click on the **Actions** tab to see how GitHub Actions is going to execute these commands:

Figure 11.9 – Failed build

As you've noticed with the red sign, the build has failed. You could click on the failed build to know more:

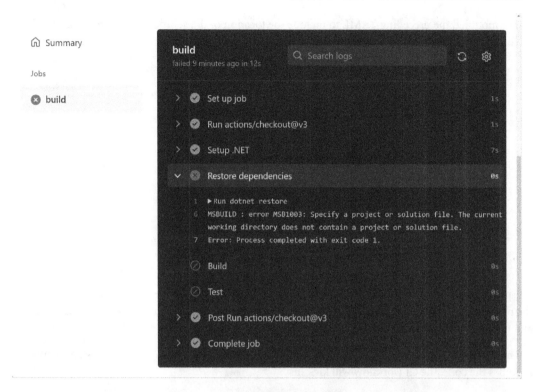

Figure 11.10 – Description of build failure

You can see why this failed: it was unable to find the solution file to restore the dependencies. This is an expected error as our solution file for *Chapter 10* lives in the /ch10/UqsAppointmentBooking directory and not on / (the root), so we need to modify the YAML file to reflect this.

Pull the latest from source control, and you will notice that you have a new directory called workflows (/.github/workflows) and inside this directory, you will find the file that we have just created: dotnet.yml.

You can edit this YAML file using any plain text editor. I use **Visual Studio Code** for this. We need to edit this file to instruct Actions where to find the solution file:

```
...
runs-on: ubuntu-latest
defaults:
  run:
    working-directory: ./ch10/UqsAppointmentBooking
steps:
...
```

I have modified the preceding YAML to include the location of the solution file. If you push this file to GitHub, this will trigger a build. You can see the results in the **Actions** tab:

Figure 11.11 – Build pass

The passing build means that all the steps that we specified in our YAML file have passed. You can click the passed build to look at how every step is executed and to ensure that your steps passed accurately and not by coincidence.

We have created our CI system on GitHub Actions for our project. Now, every time a team member modifies the code, the CI workflow will kick in.

Let's focus a bit more on the test part.

CI and testing

The last line in the YAML steps is meant to trigger all available tests in the solution:

```
- name: Test
  run: dotnet test --no-build --verbosity normal
```

This targets all the tests in your solution, which might be a combination of unit tests, Sintegration tests, integration tests, and system tests. Executing all the tests that you have might not be ideal based on what your project does. This will delay the feedback of the recent push and block other team members from validating that the build is safe to use.

So, you might want to restrict your workflow to unit tests only. You can modify the run command to the following:

```
dotnet test --filter FullyQualifiedName~Tests.Unit
    --no-build --verbosity normal
```

This will target all tests with `Tests.Unit` in their namespace. So, you don't execute all tests and you get fast feedback.

For other tests, you can write another YAML file that contains a new workflow and schedule it to run at a different event, such as multiple times per day.

Let's see what happens when a test fails.

Simulating failed tests

Let's say that a colleague forgot to run the unit tests before pushing to source control, heaven forbid. We can simulate this by changing the line from *Chapter 10's* `SlotService` from `||` to `&&` and pushing the code to source control:

```
var employeeAppointments = await _appointmentRepository.GetAppointmentsByEmployeeIdAsync(employeeId);
var appointments = employeeAppointments.Where(x =>
    x.Ending < appointmentsMaxDay &&
    ((x.Starting <= _now && x.Ending > _now) || x.Starting > _now)).ToArray();
```

Figure 11.12 – Breaking the logic in the code to trigger failed tests

This change will fail the existing unit tests in the solution, and it is usually trapped before pushing to source control, but if pushed, the CI system will report the following:

Figure 11.13 – Failed tests in source control

The previous screen will show multiple failing points and if you scroll down, it will show you more of a description of the failed test, what was expected, and what was produced instead:

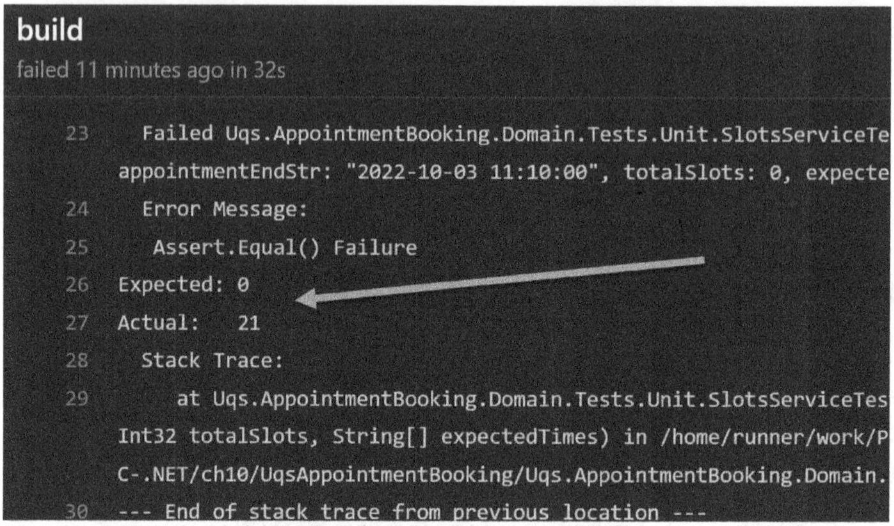

build
failed 11 minutes ago in 32s
23 Failed Uqs.AppointmentBooking.Domain.Tests.Unit.SlotsServiceTe
appointmentEndStr: "2022-10-03 11:10:00", totalSlots: 0, expecte
24 Error Message:
25 Assert.Equal() Failure
26 Expected: 0
27 Actual: 21
28 Stack Trace:
29 at Uqs.AppointmentBooking.Domain.Tests.Unit.SlotsServiceTes
Int32 totalSlots, String[] expectedTimes) in /home/runner/work/P
C-.NET/ch10/UqsAppointmentBooking/Uqs.AppointmentBooking.Domain.
30 --- End of stack trace from previous location ---

Figure 11.14 – Failed tests description

You can conclude what went wrong by reading the description. Obviously, we know what went wrong here because we broke the code on purpose. In other situations, we could understand what went wrong from the description of the error on GitHub Actions or we should be able to run the unit tests again on our local machine and try to figure out the situation.

You've seen an implementation that uses a workflow. Let's next understand the concept behind it.

Understanding workflows

The module in GitHub Actions is a workflow and a workflow looks like this:

GitHub Actions
Workflow 1
file1.yml
Event
Steps
Workflow N
fileN.yml
Event
Steps

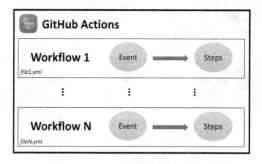

Figure 11.15 – Workflows

A workflow lives in a YAML file in the `/.github/workflows` directory in your repository. Each workflow will run when triggered by an event(s) in your repository, or they can be triggered manually or at a defined schedule. In the previous example, the events that would trigger the workflow are pushing into the main branch and raising pull requests to `main`.

I hope this section did a good job of introducing you to GitHub Actions. Of course, there are more advanced options such as `matrix` and other features, but *Rome wasn't built in a day*. You can start with the basics and progress into an expert quickly to gain fine control within your release pipeline.

Summary

A CI system is a must in today's software engineering process and a continuation of the effort that was put into TDD.

This chapter introduced GitHub Actions as a good example of a CI system and used the code of *Chapter 10* as a realistic example. By completing this chapter, I trust you should be able to add CI configuration to your arsenal of tools.

In the next chapter, we will see how to consider adding tests to a brownfield project, as we don't always have the luxury of starting a greenfield project all the time.

Further reading

To learn more about the topics discussed in the chapter, you can refer to the following links:

- *YAML files*: `https://yaml.org/`
- *GitHub Actions workflows*: `https://docs.github.com/en/actions/using-workflows/about-workflows`
- *Continuous Integration* by Martin Fowler: `https://martinfowler.com/articles/continuousIntegration.html`

12

Dealing with Brownfield Projects

I cringe whenever I hear **brownfield project**, and probably you do too. Design decisions are already taken, code is already written by previous developers, and code quality varies between one class and another; brownfields are not for the faint-hearted.

As there could be multiple definitions for the term brownfield, I want to start by defining it here, so we are all on the same page. From this book's perspective, a brownfield project is a project not covered by unit tests and was probably written a while ago. It might have been covered by other types of tests than unit tests, but we will still refer to it as a brownfield. Some techies also refer to it as a **legacy project**.

As you have already figured out, we have dedicated a whole chapter to brownfields as there are challenges in introducing TDD or unit testing to such projects. We will discuss those challenges and how to overcome them.

In this chapter, we will cover the following topics:

- Analyzing the challenges
- The strategy for enabling TDD
- Refactoring for unit testing

By the end of this chapter, you will better understand what you need to look for when enabling unit testing for your project. You will also get an insight into required code changes.

Technical requirements

The code for this chapter can be found at the following GitHub repository:

```
https://github.com/PacktPublishing/Pragmatic-Test-Driven-Development-
in-C-Sharp-and-.NET/tree/main/ch12
```

Analyzing the challenges

In the previous chapters, we've been talking about adding new features while starting from the unit tests end (testing first). We relied on having a new functionality or modifying an existing functionality that is already covered by existing unit tests. This is not the case for brownfields as, when trying to apply TDD, you will face some of these challenges:

- **Dependency injection support**: Some legacy frameworks do not natively support DI, which is necessary for unit tests.

- **Code modification challenges**: Changes to code that are not covered by tests (of any type) can introduce new bugs.

- **Time and effort challenges**: Introducing the ability to unit test the code requires time and effort.

Let's go through each challenge in detail, so you can consider them when the time comes.

Dependency injection support

In this book, before learning about unit testing or TDD, we had to introduce DI. DI is what allows you to separate your code into units/components; it is a natural requirement for unit testing. There are two challenges in enabling DI – framework support and refactoring work. Let's dig deeper.

Framework support for DI

This is a .NET book, so we are only interested in legacy .NET frameworks that do not support DI natively. In the early 2000s, while unit testing was becoming a trend, Microsoft was more interested in migrating developers from **Visual Basic 6 (VB6)** and **Active Server Pages (ASP)**, thus enabling native DI in the early days of .NET was not on the priority list.

Therefore, Win Forms and ASP.NET Web Forms were born with no native DI support. Surely, you can hack the framework and add some support for DI. Still, when you start shifting from the norms of a framework, you alienate other developers working on the code base and introduce subtle bugs and complexity to the design.

More modern frameworks, such as WPF, and ASP.NET MVC from the classical .NET Framework allowed dependency injection through third-party DI containers. Today, with ASP.NET Core, DI is natively supported using Microsoft-built DI containers.

If you have a project that is built on a legacy framework with no native support for DI, such as Win Forms and ASP.NET Web Forms, I would say the effort put into bending these frameworks to enable unit testing needs to be weighed against the benefits of having unit tests. Maybe you can invest this effort in applying other types of testing to the project instead. Obviously, migrating the project into a modern framework solves this problem, but that has its own challenges as well.

If the framework supports DI natively or can support DI with little effort, then you are in luck, but is that all? Clearly, now you have to refactor everything to enable DI.

Refactoring for DI support

We dedicated *Chapter 2, Understanding Dependency Injection by Example,* to discussing DI, so we will not go into details here. What we need to do when we plan to introduce unit tests or TDD is to make sure we are using DI to inject components.

Ideally, all your components need to be injected through constructor injection, and instantiating a variable should not be done in the method or property code. With that said, consider the following unwelcome code:

```
MyComponent component = new MyComponent();
```

When you have code that is not unit tested, you'll probably find that all the components are instantiated in the code, and no DI container is used. In this case, you will have to go through the cases of direct instantiation and modify them to support DI. We will see an example of this at the end of this chapter.

Not all cases of direct instantiation will require you to refactor them for DI. Some cases are part of a standard library that you could unit test, but you shouldn't. Take this line as an example:

```
var uriBuilder = new UriBuilder(url);
```

In this example, we had no intention of injecting the `UriBuilder` class, so you might not need to change the code because the class is not relying on an external dependency. Accordingly, injecting the class is not beneficial, but actually, it adds a bit of unnecessary effort.

In brief, to make the code unit testable, all the components need to be DI-ready. Depending on how big your project is and the way you want to implement it (such as iteratively), it will take time and effort.

Introducing DI is not the only challenge; modifying the code will pose a new challenge as well.

Code modification challenges

When you are adding non-unit tests to your project, you are working externally to the code and some of the activities can be as follows:

- Testing the UI with an automation tool such as Selenium or Cypress. The tests will deal with the application like an external user.
- Doing integration testing by performing an end-to-end call, say on an API endpoint.
- Load testing a project by creating multiple instances of the application.
- Penetration testing by trying to hack into the application externally.

All these activities do not require changing the code, but unit testing requires the production code to be in a certain shape.

When changing the code to enable unit testing, we risk breaking it. Imagine having a bug that found its way to production and the irony of the business hearing *it broke because we added tests*.

> **Important note**
> I would be lying if I said that you have to change the code for unit testing because you can use an isolation testing framework, which would allow you to unit test the code without changing it. However, this would be the last resort if you really want to unit test and you cannot change the code. We'll discuss this further later in this chapter.

There are solutions to change the code that will reduce the possibility of breaking, so follow along.

Time and effort challenges

The process of enabling dependency injection and refactoring the code into components that act as units is mentally challenging and time-consuming.

Think of approaching the process iteratively by dividing it into your sprints (or iterations, or whatever you call them) or by blocking some iterations and implementing your changes.

The challenge here is to justify to the business the time spent in introducing unit testing and enabling TDD, as from their perspective, you still have the same product, the same number of bugs, and nothing is fixed, but you have just added tests. Obviously, you and I know that unit testing will guard the code against future bugs and add documentation, but the challenge is to convey this to the business. The next chapter is going to tackle dealing with the business when introducing TDD and unit testing, so I will stop here.

All these challenges have solutions; after all, we work in software! The next sections will address them with different strategies.

The strategy of enabling TDD

Now is the time to discuss solutions to the challenges described in the previous section. Since *a picture is worth a thousand words*, I will present a workflow diagram that should clarify how to introduce unit testing into a brownfield project:

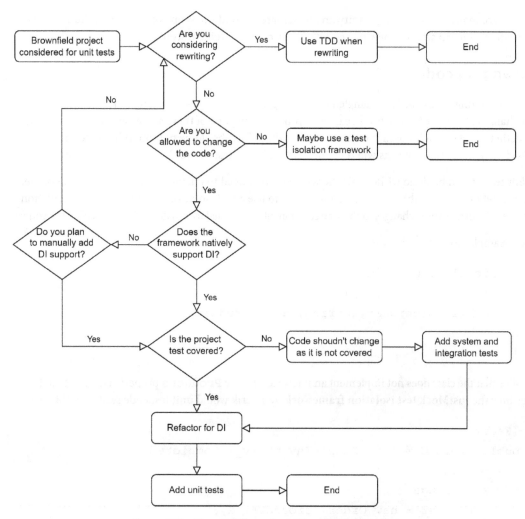

Figure 12.1 – The workflow of enabling TDD in a project

Let's go through the diagram and our options.

Consider rewriting

You might consider rewriting because the existing project might be based on an old framework with fewer developers and less support. However, the rewriting idea is controversial. If you tell the business that the project requires rewriting, you will be their least favorite individual. Trust me; nobody wants to hear this. However, good rewriting doesn't need to be a big bang; it can be divided into smaller chunks of upgrades and can be appended to sprints. Obviously, choosing a modern framework that supports DI natively or through a third party is out of the question.

There are many ways to rewrite software, which are beyond the scope of this book. But if you are rewriting, you can start the new pieces with TDD, and the problem would be sorted!

Changing code

In some settings, the code is too tangled to be changed, or sometimes the business doesn't like the idea of changing the code for whatever reason. If you are facing one of these situations, then ask yourself whether it is worth adding unit tests or whether this effort should be put into other types of testing. Obviously, other types of tests will do good, although unit testing would even be more beneficial.

Unit testing can be done without DI; hence, you don't need to change the code. Here you go, I let the secret out! But for this to work, you will have to use a **test isolation framework**. A test isolation framework makes some changes to the way components are loaded externally without touching the code.

For example, consider this class:

```
public class Warehouse
{
    public Dictionary<string, int> Products { get; }
    ...
}
```

Notice that the class does not implement an interface, and the Products property is not virtual. Let's see how the **JustMock test isolation framework** by Telerik would unit test code related to this class:

```
[Fact]
public void Complete_SampleInventory_IsCompleted()
{
    // Arrange
    var order = new Order("trouser", 1);
    var warehouse = new Warehouse();
    Mock.Arrange(() => warehouse.Products)
        .Returns(new Dictionary<string, int>() {
            { "shirt", 12},
            { "trouser", 5}
    });

    // Act
    order.Complete(warehouse);
```

```
    // Assert
    Assert.True(order.IsCompleted);
}
```

In this code block, we only care about the two highlighted lines. **JustMock** was able to mock the Products property on the Warehouse class, although the property is not virtual and the warehouse object is not instantiated with the help of the mock library.

JustMock did some magic here, it made the Warehouse class mockable, despite that Warehouse doesn't have an interface and Products is not virtual. No DI was required!

However, the isolation framework's magic is not liked by most TDD practitioners, as it leads to bad programming practices. Plus, these frameworks are not free. They definitely solve a problem when you want to avoid code changes, but they pose the question, is it worth the effort and the cost?

Going through the hassle of having a dependency on a non-standard way of testing will require training, maintenance, and licensing costs, which should be weighed against utilizing any framework.

Native support for DI

Some frameworks within .NET have no notion of DI – Win Forms and Web Forms are perfect examples. You can force them to support DI, but this means bending the framework and being on your own. Sometimes, you can try to isolate the UI layer and unit test what is underneath it. That is good enough in this case.

What I want to say is having a framework that doesn't natively support plugging in a DI container or has one that is built in, such as ASP.NET Core, will cost you more effort and will shift you away from the norm.

I would avoid unit testing such frameworks and promote the quality by employing other testing categories.

Test coverage prior to unit test

Code changes will lead to bugs, but what about careful code changes? Well, yes, it will still lead to bugs! Bugs will come when you change code, no matter how careful you are. So, what is your bug hunting plan?

If you plan to change the code for unit testing, your code should have high coverage at first by other types of tests, mainly automation and integration. These tests will help to point out where you've broken the code before it gets to production.

The logical question coming up is, if I have high coverage with other types of tests, why do I need unit tests? The answers are as follows:

- If your project is still in development, then you will need unit tests. Also, preferably add new features in TDD style.

- You can change the balance of all the available tests to unit tests once your project can support them, as unit tests have advantages over other tests that we've discussed earlier in this book, in *Chapter 4, Real Unit Testing with Test Doubles.*

If your code is in maintenance mode and the coverage is already high, then I would argue that adding unit tests is not very useful. In this case, TDD would not be applicable as TDD is a companion of new features or features change.

My advice is not to change the code if it is not covered by tests, as your valuable effort in advancing the project might be countered by production bugs. Maybe the effort should be put into other tests or rewriting.

Every project is different and the strategies we have mentioned here are just points to consider. You should consider adding these to your train of thought when you plan the introduction of unit tests into a brownfield.

Next, we will see samples of changing legacy code to allow unit testing.

Refactoring for unit testing

When you write in TDD, your code is unit testable from the first moment. This is because you took into consideration DI scenarios. Brownfield code almost always has no consideration for DI, and it will have to change to accommodate it.

In this section, we will cover the scenarios that you have to change, and then we will go through an example of a sample refactoring at the end of this section.

Variables instantiated in the code

Whenever you see a new keyword in the code that is instantiating a library or a service, then most probably, this needs refactoring. Take the following example of code in a method:

```
var obj = new Foo();
obj.DoBar();
```

The previous line means we cannot inject a test double for Foo, so the code needs to change to inject it.

The next thing to do is to check whether Foo implements an interface for the methods you are using from this class. Let me break the bad news for you here – keep your expectations low; you most probably won't find that the class implements an interface for the methods you are using unless you are consuming a well-designed and sophisticated framework.

In the next few sections, we will go through the process of making the code testable.

Creating an interface for your own class

If you own the code in Foo and you can change it, great! Your code can change from:

```
class Foo
{
    public void DoBar();
}
```

To this class and an additional interface, IFoo:

```
interface IFoo
{
    void DoBar();
}
class Foo : IFoo
{
    public void DoBar();
}
```

This is easy. But what if the source code of this class is not accessible to you, or you are not allowed to change the source code?

Creating an interface for a third-party class

Adding an interface for a class you do not own is not possible. You have to go through another pattern, usually referred to as a wrapper class. You will need to create a new class and interface, as such:

```
interface IFooWrapper
{
    void DoBar();
}
class FooWrapper : IFooWrapper
{
    private Foo _foo = new();
    public void DoBar() => _foo.DoBar();
}
```

You can see that we have wrapped the Foo class with another class to intercept the calls to the DoBar method. This would allow us to add an interface in the same way we can add an interface to a class we own.

There is a bit of extra work here, but you will get used to it, and it will become straightforward after a couple of class changes.

Now that we have an interface for our class, we can go to the second step, DI.

Injecting your components

How you do dependency injection depends on the library you are using (ASP.NET Core, Win Forms, among others) and the way you've wired up your DI container. Let's take an ASP.NET Core WebAPI project. To wire up your newly created or updated class, write code similar to the following in Program.cs:

```
builder.Services.AddScoped<IFoo, Foo>();
```

Or the following code:

```
builder.Services.AddScoped<IFooWrapper, FooWrapper>();
```

Obviously, the lifespan scope (transient, scoped, or singleton) will change based on the Foo class.

Once you've done the modification, you can refactor your controller to inject FooWrapper:

```
public class MyService
{
    private readonly IFooWrapper _foo;
    public MyService(IFooWrapper foo)
    {
        _foo = foo;
    }
    public void BarIt()
    {
        _foo.DoBar();
    }
}
```

We introduced a wrapper class and an interface so we can follow along with a familiar DI pattern, so the previous code became possible.

Now, you can go ahead and implement whatever unit testing you want to put in place, as you can inject a test double for `FooWrapper` at test time.

The instantiation scenario is sorted. Let's explore another refactoring pattern.

Static members replacement

Static methods, which include extension methods, are simple, occupy fewer coding lines, and produce beautiful code. However, they are evil when it comes to dependency injection; as per the explanation in *Chapter 2, Understanding Dependency Injection by Example*, static methods are not unit test friendly.

`Date.Now` looks innocent, and Now is a read-only static property. If you want your unit test to freeze the time, for example, say you want to test what happens on February 29 (leap year), you can't do that. The solution to this one is a wrapper, as discussed earlier. This is what you can do to make Now an instance method rather than a static method:

```
public interface IDateTimeWrapper
{
    DateTime Now { get; }
}
public class DateTimeWrapper : IDateTimeWrapper
{
    public DateTime Now => DateTime.Now;
}
```

We've done exactly what we've done earlier where we did not have control over the class (few sections back). We enabled DI support by introducing the wrapper pattern to the `DateTime` class. Now, you can inject `DateTimeWrapper` at runtime and use a test double for unit testing.

If you have control over the class, you might want to change the static member to an instance one (non-static) or introduce an additional instance member and keep the static member:

```
Interface IFoo
{
    string PropWrapper { get; }
}
class Foo : IFoo
{
    public static string Prop => …
    public string PropWrapper => Foo.Prop;
}
```

This is a way to expose your static property as an instance property. You will also have to use the `PropWrapper` wrapper property instead of the non-wrapped one, `Prop`, in the rest of your code. In the previous example, we added an additional property, but you could also refactor the code to replace the static property if it makes sense.

Changing the consumer to rely on an instance member

The code consuming the previous `Foo` class might look like this:

```
public class Consumer
{
    public void Bar()
    {
        ...
        var baz = Foo.Prop;
        ...
    }
}
```

After refactoring `Foo` as per the previous section, the implementation here can change into a unit-testable format, as follows:

```
public class Consumer
{
    private readonly IFoo _foo;
    public Consumer(IFoo foo)
    {
        _foo = foo;
    }
    public void Bar()
    {
        ...
        var baz = _foo.PropWrapper;
        ...
    }
}
```

You can see that we have injected `IFoo` into the `Consumer` class and we have used another property, `PropWrapper`.

Instantiated classes and static member calls can be easily spotted. However, the most notable thing about legacy code is that it doesn't have a structure, and a component cannot be easily noticed and tested. So for this, we will have to make more changes.

Changing code structure

The code in the brownfield project might be in a format that isn't unit testable. One popular structure is the controller's action method with all the code written inside it:

```
public void Post()
{
    // plenty of code lines
}
```

Here, we need to get the code into a unit-testable structure. I would choose an architecture such as DDD, as per *Part 2* of this book, where we've used services and domain objects.

The previous sample code works fine, but it is not unit testable. You can find the full listing in the `WeatherForecasterBefore` directory of this chapter in the `WeatherForecastController.cs` GitHub file:

```
public class WeatherForecastController : ControllerBase
{
    public async Task<IEnumerable<WeatherForecast>>
    GetReal([FromQuery]decimal lat, [FromQuery]decimal lon)
    {
        var res = (await OneCallAsync(lat, lon)).ToArray();
        ...
        for (int i = 0; i < wfs.Length; i++)
        {
            ...
            wf.Summary = MapFeelToTemp(wf.TemperatureC);
        }
        return wfs;
    }
    private static async
        Task<IEnumerable<(DateTime,decimal)>> OneCallAsync(
        decimal latitude, decimal longitude)
    {
        var uriBuilder = new UriBuilder(
```

```
            "https://api.openweathermap.org/data/2.5/onecall");

        ...

        var httpClient = new HttpClient();

    }
    private static string MapFeelToTemp(int temperatureC)
    {

        ...

    }
}
```

Clearly, most of the code is omitted for brevity. The code will call a third-party service called Open Weather, and get the forecast for the next 5 days for a certain geographical coordinate. It will then analyze the temperature and produce a word that describes the temperature feel, such as *Freezing* or *Balmy*.

The previous code also instantiates a `HttpClient` instance, which means there is no easy way of avoiding calling the third party whenever we try to unit test this.

Next, we shall invest some thoughts into changing this code into testable components.

Analyzing code change for a testable format

The code that we've just seen can be made into components in several ways, and there is no one way to do it. This code does two things, so we can think of two components that will encompass all the code functionality:

- Calling Open Weather and obtaining a forecast
- Obtaining the forecast and analyzing it

The idea here is to have the controller with no business logic and if there is no business logic, then we will not need to unit test the controller. The controller, in general, should have no business logic, and it should do a single job – deliver **data transfer objects** (**DTOs**) to views (as of Model-View-Controller views).

We will give the following names to our components:

- `OpenWeatherService`
- `WeatherAnalysisService`

The whole call to obtain the forecast and the temperature feeling analysis would look like this:

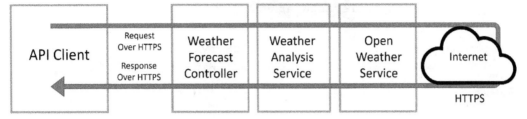

Figure 12.2 – Workflow by components

The client will call the APIs to obtain the forecast with the feeling. The weather forecast controller will receive the call and pass it to the weather analysis service, which loads the Open Weather service and calls an external dependency to obtain the weather.

Next, we will see what the code looks like after our refactoring.

The final testable code

There are levels of refactoring invasiveness when you want to enable unit testing. I chose an aggressive level, but you might choose to refactor less code.

You can see the whole refactored code in the **WeatherForecasterAfter** directory.

Now the controller looks like this:

```
public class WeatherForecastController : ControllerBase
{
    private readonly IWeatherAnalysisService
        _weatherAnalysisService;
    public WeatherForecastController(
        IWeatherAnalysisService weatherAnalysisService)
    {
        _weatherAnalysisService = weatherAnalysisService;
    }
    [HttpGet]
    public async Task<IEnumerable<WeatherForecast>>
        GetReal(
        [FromQuery]decimal? lat, [FromQuery]decimal? lon)
    {
        if (lat is null || lon is null)
        {
```

```
            return await _weatherAnalysisService
                .GetForecastWeatherAnalysis();
        }
        return await _weatherAnalysisService
            .GetForecastWeatherAnalysis(lat.Value, lon.Value);
        }
    }
```

The controller is almost empty, compared to how it was before. The action method in the controller is mapping the API call to the right service.

This is the OpenWeatherService class:

```
public class WeatherAnalysisService :
    IWeatherAnalysisService
{
    ...
    private readonly IopenWeatherService
        _openWeatherService;
    public WeatherAnalysisService(
        IOpenWeatherService openWeatherService)
    {
        _openWeatherService = openWeatherService;
    }
    public async Task<IEnumerable<WeatherForecast>>
        GetForecastWeatherAnalysis(decimal lat, decimal lon)
    {
        OneCallResponse res = await
            _openWeatherService.OneCallAsync(...)
        ...
    }
    private static string MapFeelToTemp(int temperatureC)
    {
        ...
    }
}
```

The class contains the logic of mapping the feeling to the temperature and calling OpenWeatherService. The service doesn't know how the Open Weather APIs are called.

Finally, let's look at OpenWeatherService:

```
public class OpenWeatherService : IOpenWeatherService
{
    ...
    public OpenWeatherService(string apiKey,
        HttpClient httpClient)
    {
        _apiKey = apiKey;
        _httpClient = httpClient;
    }
    public async Task<OneCallResponse> OneCallAsync(
        decimal latitude, decimal longitude,
        IEnumerable<Excludes> excludes, Units unit)
    {
        ...
    }
}
```

The job of this service is to wrap the HTTP call to the internet Open Weather APIs with code.

The full unit tests for the newly added services can be found in the source code directory with the rest of the services.

Remember that we are doing the refactoring while assuming the code has undergone other types of testing. Aggressively refactoring code is time-consuming, especially for the first set of refactors, but remember that refactoring is also paying back part of the technical debt of the project. The code is now documented by the unit test as well, so this is a step forward.

Summary

In this chapter, we discussed the implications of enabling unit testing for a brownfield project. We've gone through the considerations to allow you to decide whether it is worth it and all the things that you have to watch for while progressing with the process.

As a developer, you will encounter brownfield projects that went up in value and would benefit from unit tests and TDD. Hopefully, this chapter equipped you with the required knowledge to tackle them.

Deciding to introduce TDD into your organization is not a straightforward process. The next chapter will go through the process and prepare you for some scenarios that you would face.

Further reading

To learn more about the topics discussed in the chapter, you can refer to the following links:

- *JustMock (Isolation Framework from Telerik)*: `https://docs.telerik.com/devtools/justmock`

- *Microsoft Fakes (Isolation Framework that comes with VS Enterprise)*: `https://docs.microsoft.com/en-us/visualstudio/test/isolating-code-under-test-with-microsoft-fakes`

- *TypeMock (Isolation Framework for classical .NET Framework)*: `https://www.typemock.com`

13

The Intricacies of Rolling Out TDD

I have frequently seen developers putting their efforts into trying to convince the business to follow TDD or adopt unit tests. In fact, this is a situation in which I have often found myself, and for this reason, I want to share my experience with you in this chapter.

After reading this book, you might feel strongly about implementing TDD in your direct team or large organization to reap the quality benefits. So far, so good. The second stage is doing this in a structured manner and being prepared for the business' counter-arguments and rejections.

We will highlight the challenges and guide you through the process of convincing your business and team to take the TDD approach. In this chapter, we will discuss the following topics:

- Technical challenges
- Team challenges
- Business challenges
- TDD arguments and misconceptions

After reading this chapter, you will be ready to present your team and/or business with a plan to move forward with TDD.

Technical challenges

There is a set of technical and business challenges an organization must overcome before adopting TDD. Here, we will cover the technical challenges, and in the next section, we will consider the team challenges and then the wider organization challenges (business challenges). We will start with a diagram to explain the workflow of rolling out TDD in your organization:

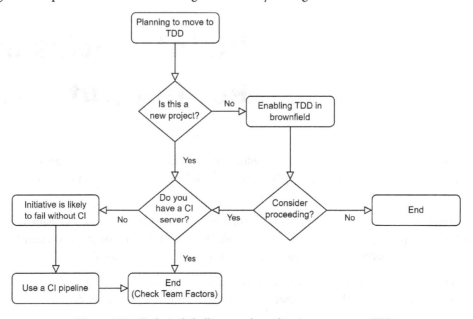

Figure 13.1 – Technical challenges when planning to move to TDD

We will go through the diagram in the next sub-sections, so let's start.

Greenfield or brownfield?

If you are working on a brownfield project, the technical challenges were well presented in the previous chapter, so I will not go further into these challenges. To introduce TDD, you need to consider the effort, suitability, and alternatives.

If you are starting a new project (a greenfield project), then you are in luck. You can go ahead with your plan.

Tools and infrastructure

Today, with the availability of the cloud, having an infrastructure to run your **continuous integration** (**CI**) pipeline is easy and cheap. However, some organizations have restrictions on using the cloud and you might struggle to get a CI server.

If you don't have a CI server in place, then at the risk of sounding pessimistic, doing TDD is doomed to fail. This is because developers will break the unit tests and you will have them disabled or failing.

Some developers also like to invest in tools such as **JetBrains ReSharper** for its good-quality test runner and refactoring capabilities, but this is optional. Also, you may want to consider JetBrains Rider as it has all the capabilities of ReSharper, as discussed in *Chapter 1* of this book.

However, if you are using MS Visual Studio Professional 2022 or later, you already have a good tool for a proper TDD process.

The technical challenges are not all you need to think about. Consider also your team and its readiness to embrace TDD and then your business challenges. Let's continue with the team challenges.

Team challenges

If you are a solo developer working on a project, no worries, you can do whatever. However, most business projects are implemented by a team, so making the effort to use TDD is a team decision. Again, let's start with a workflow diagram:

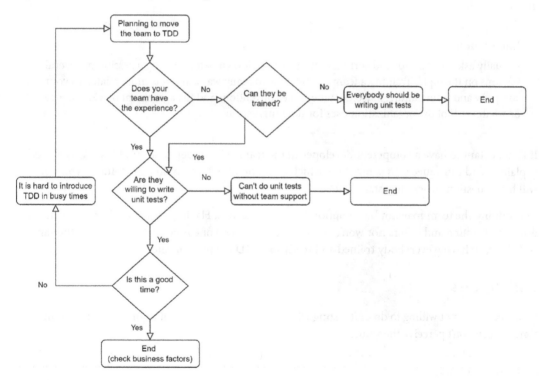

Figure 13.2 – Team challenges when planning to move to TDD

We will go through this diagram in the next sub-sections. Let's go through the points to keep in mind when planning to move your team – whether you are a developer wanting to influence the team or in a position where you can enforce technical standards.

Team experience

Unit testing requires DI, which in turn requires experience in OOP. Your team members may be unfamiliar with unit testing or may mistake unit testing with integration testing.

> **Important notes**
>
> The xUnit and NUnit libraries are widely used to implement integration tests. Because they have the suffix **Unit**, developers sometimes incorrectly assume the written tests are unit tests. I've seen teams claiming they have unit tests, but when I inspected the code, I discovered otherwise.

If your team requires training in TDD, then they need to be educated on what it is, how to do it, and the value of TDD. I have a book in mind to recommend for the training, but I will leave it to you to guess which it is.

> **Important notes**
>
> I usually ask the team to read certain material on their own before having a session or several sessions on the topic. Training a team can be done in many ways, and it is more related to your company and team culture. I also document the conventions and the agreement on Confluence or whatever tool the organization uses for documentation.

It is important to have a competent developer in the team, who understands unit testing, is a good explainer, and can squeeze in some time, which could be you. This would be helpful, as your team will have questions when they start TDD.

But training the team may not be an option for many reasons. Having a few members of the team doing unit testing and others not won't be productive, as everyone is going to operate on the same code base, so having everybody trained and ready for TDD is a prerequisite.

Willingness

Some teams are not willing to do unit testing whether they feel it is difficult or increases development time or they don't perceive the value.

> **Important notes**
>
> I have seen unit testing enforced by an organization, but the team was unwilling to write the tests, and they just created a unit test project with meaningless tests to tick the box of the question *Have you implemented unit testing?*.

Having the team synchronized with the objective and collaborating for one goal is valuable in promoting product quality.

If your team is not willing to go for TDD for whatever reason but unit testing is OK, then go for it! You can gently introduce TDD soon after. It doesn't need to be all or nothing. It is also worth noting that some members can do TDD while others can do unit tests.

Unit testing is useless

I have heard this argument from many developers. They might have formed their opinion on bad implementations of unit testing or have other reasons. Certainly, unit testing has some disadvantages, but so do most technologies.

Your best shot is to understand the reasons behind this misconception and see whether you can address them.

TDD is useless, I would do unit testing

TDD is controversial and sometimes developers have their own experiences that told them that TDD is unusable. This is OK as long as they are happy with unit testing because not all team members have to do TDD.

If the developers who don't believe in unit tests are building their arguments on seeing bad practices, then your job might be to promote good practices.

The team's willingness to follow TDD has a crucial impact on the success of the project, so it is important to have everybody on the same page.

Timing

TDD requires some preparations and extra effort to get the essential quality, following the *no pain, no gain* mantra. Having the right time is important and it should definitely not be near the release time or when the team is stressed.

The perfect time is at the start of the project, but there is no harm in introducing it later.

Once you've passed the first and second challenges, you will have the business challenges, which are arguably the hardest.

Business challenges

The business here means a higher technical authority outside the team, who can enforce rules. Also, it can be the project manager or the product owner.

I believe that a successful rollout of TDD or unit testing comes from top to bottom, management-wise. Enforcement can come from:

- Head of development
- Development manager
- Team lead
- Technical lead
- IT auditing

If this is a personal initiative or a team initiative, the team might think of dropping it under delivery pressure. However, if they are responsible for providing unit tests as part of the delivery, including a coverage level, then it cannot be missed.

Let's think of TDD from the business perspective, so that we are better equipped and articulated in getting our points across.

Business benefits of TDD

We are well aware of what the benefits of TDD are from a technical point of view. But businesses would be more open to the benefits from the business point of view, so let's get into it.

Fewer bugs

This is clearly the biggest seller, as nobody likes bugs. Some businesses are stung by a high number of defects in their products, and having less is definitely a welcome promise.

The only issue is that it is hard to prove a lower number of bugs via statistics – the project will have unit tests from day one, so we can't make a comparison between before and after.

Live documentation for the project

One thing that worries the business is documentation, which has a tight relationship with developers' turnover. The risk is that if a developer leaves, some of the business knowledge is lost. To prevent this situation, it is important to have the business rules robustly documented, and, in all honesty, I can't think of any tool that is better for this than unit tests.

Project documentation contains documents that cannot be covered by unit tests, such as project architecture. However, the detailed business rules that nobody will remember in a few months will be covered by unit tests and monitored with every developer source control push.

Promoting unit testing as a documentation tool is powerful and will give you listening ears from the business.

Fewer testing resources

In the old days, manual testing took a good chunk of the **software development life cycle** (SDLC). Today, with unit tests and other automated tests, manual testing has shrunk in size and the number of required manual testers has gone down. Some organizations have even eliminated manual testing completely in favor of having automated tests (including unit tests).

So, the promise of unit tests is about covering lots of edge cases and business rules with fewer testers and almost no regression test time.

> **Important note**
> **Regression testing** is going through existing functionality to ensure it is still working. This typically happens before a new release.

Obviously, fewer testing resources means less cost, while less time means shipping a feature faster, which naturally leads us to the next topic.

Ability to release in short cycles

The development model today in more agile organizations has moved to shipping a few features now and then.

Having unit tests regressing the code with every change and a CI/CD system in place means your software is ready to be shipped anytime.

No clever business person will believe all these aforementioned benefits will come at no cost, so next, we will discuss the disadvantages of unit testing.

Disadvantages from the business perspective

In general, extra quality requires more effort, and TDD is no different, but luckily the disadvantages are minimal.

Slight delay in the first release

We have discussed before that teams not using TDD tend to deliver faster in the first period; we have spoken about this in *Chapter 5, Test-Driven Development Explained*. Here is a quick reminder:

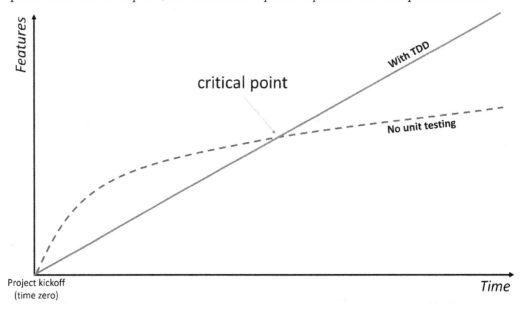

Figure 13.3 – TDD versus no unit testing

The idea here is that the effort of writing unit tests adds to the development time in the short term, but the speed becomes faster in the mid and long terms.

This is a small price to pay for quality, but just be aware of this point.

Delay in the first release is not acceptable

There are circumstances where the business wants the first version as soon as possible and they are not interested in looking beyond this. Here are a few scenarios:

- The product manager wants the first version out as soon as possible, as this can lead them to get a higher bonus or a promotion.

- There is a competitive advantage in releasing as soon as possible and not worrying about the future. This is the mentality of a start-up trying to survive.

- This project is done for a third party, and the business doesn't get paid extra for ensuring added quality. But getting this in the shortest time possible is the intention.

It would be clear if the business is not interested, and you can feel this in advance if you know the business model that the company uses. This is not a criticism against TDD, but it would become a disadvantage when mixed with such scenarios.

Now that we have gone through all the challenges and the merits of TDD, let's form a plan to introduce TDD.

TDD arguments and misconceptions

Here are a few hints and tips – from my own experience – that will occur again and again in a conversation with the business or your colleagues.

Unit testing, not TDD

When discussing with the business, to reduce the complexity of the conversation, especially if the business is not tech-savvy, use the term unit testing rather than TDD. TDD is a technical process that individuals will do themselves and it is not directly related to the business, so why complicate the discussion by adding it? Sometimes the business has heard the term TDD, and they are excited about it, so then TDD is the right term to use!

My advice is to use unit testing in your conversation unless the business has some preference for the term TDD.

Unit testing is not implemented by testers

The term *testing* in unit testing is misleading to the non-techies as it implies a tester doing manual testing. I have had this conversation with many business individuals.

It is important to clarify that unit testing has more functionality other than just testing, such as the following:

- Shaping the code design architecture of the project
- Live documenting the code
- Instant feedback during development when breaking a business rule

Also, unit tests are written in C# (or whatever other language you are using), and they are implemented by the same developer writing the code. A manual tester, most probably, won't have the willingness or the expertise to write these tests.

This argument might be triggered when the business is wondering why you want to consume the engineers' expensive time on unit testing when there are testers that could (as they initially thought) do the unit tests.

The way for writing and maintaining documentation

I am sure experienced business people will relate to the lack of documentation or the documentation becoming out of date.

As you already know, documenting the code with unit tests provides fresh documentation versus text-based stale documentation that is written and forgotten or covers part of the system (a hit and a miss). The keyword here is up-to-date, fresh, or live. Obviously, we are talking about part of the documentation and you might have to make this clear. This is the detailed business rules part.

We have incompetent developers

Businesses sometimes believe they have incompetent developers, and this is why they are producing plenty of bugs. I have heard this argument whispered by the business on multiple occasions when talking about their team.

When I hear this argument, I quickly dig and discover that the business does not have a structure in place for the agile process and that the developers are rewarded for how fast they finish developing a feature. We all know that person who takes the shortest route to get their features done and show off to the business!

Developers are highly logical individuals who like structure and order. Having a development process with TDD will definitely reduce bugs and set things on track.

Your challenge here is to show how the TDD process and the tests will have a positive effect on the problem.

Summary

This chapter utilizes all the knowledge provided in this book and demonstrates the challenges of rolling out TDD into your organization. I hope I gave you enough arguments to convince the team and the business to subscribe to the TDD point of view.

Besides this chapter, your presentation skills and familiarity with the subject will be highly useful when planning to roll out TDD.

In this book, I have endeavored to provide practical examples of real frameworks and tools that I've worked with, rather than using abstract and oversimplified examples. I wrote the book out of love and passion for the topic and I tried to stay pragmatic, and I hope I delivered what I aimed to deliver.

While the title of this book refers to TDD, this book contains pragmatic examples of OOP and good programming practices, and by finishing the book, I trust you have stepped into the world of advanced software engineering.

Good luck and I would love to know how the book has contributed to you or your team adopting TDD.

Appendix 1: Commonly Used Libraries with Unit Tests

We have used two major libraries for unit testing across the book:

- xUnit
- NSubstitute

Your team may be using these libraries already. Or you may have done a bit of experimentation with unit testing, and you want to expand your horizon into more libraries. While these libraries are popular, other libraries can replace them or work side by side with them. This appendix will skim through the following libraries:

- MSTest
- NUnit
- Moq
- Fluent Assertions
- AutoFixture

All these libraries use the MIT license, the most permissive license and you can install any of them via NuGet.

By the end of this appendix, you will be familiar with the libraries that form the ecosystem of unit testing in .NET.

Technical requirements

The code for this chapter can be found at the following GitHub repository:

https://github.com/PacktPublishing/Pragmatic-Test-Driven-Development-in-C-Sharp-and-.NET/tree/main/appendix1

Unit testing frameworks

We have seen xUnit, and we have briefly spoken about MSTest and NUnit. This section will give you a feeling of what these other frameworks are about.

MSTest

MSTest used to be popular, as it was installed as part of **Visual Studio** (**VS**) in the older versions of VS. Prior to NuGet's existence, using a *built-in* library could cut configuration and deployment time compared to adding and using another framework such as NUnit.

Before NuGet, installing a new library involved manually copying DLLs, putting them in the right location, changing some configurations, and pushing them into source control for the team to share the same files. So, having a pre-installed library and one that didn't require configuration, such as MSTest, was a blessing. We have moved a long way since then.

To add an MSTest project into your solution, you can do it via the UI:

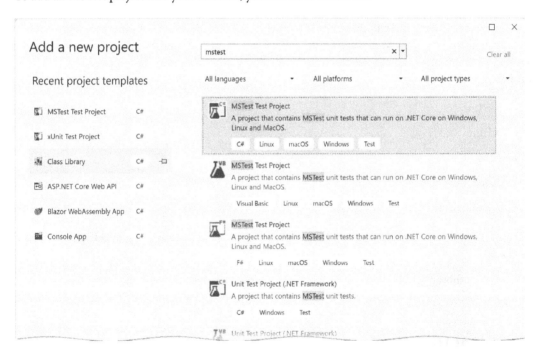

Figure A1.1 – Adding MSTest via the UI

Notice that there are two copies of the C# version. The bottom one is for the classical .NET Framework, and the top one is what we use with .NET Core.

You can add MSTest via the dotnet CLI:

```
dotnet new mstest
```

MSTest and xUnit have similar syntax, so let me show you the code in xUnit and its equivalent in MSTest. I'll start with xUnit:

```
public class WeatherAnalysisServiceTests
{

    ...

    public WeatherAnalysisServiceTests()
    {
        _sut = new (_openWeatherServiceMock);
    }


    [Fact]
    public async Task GetForecastWeatherAnalysis_
        LatAndLonPassed_ReceivedByOpenWeatherAccurately()

        ...

        // Assert
        Assert.Equal(LAT, actualLat);
        Assert.Equal(LON, actualLon);
    }
    ...
```

The equivalent in MSTest is as follows:

```
[TestClass]
public class WeatherAnalysisServiceTests
{

    ...

    [TestInitialize]
    public void TestInitialize()
    {
        _sut = new(_openWeatherServiceMock);
    }


    [TestMethod]
    public async Task GetForecastWeatherAnalysis_
```

```
        LatAndLonPassed_ReceivedByOpenWeatherAccurately()
    {

        ...

        // Assert
        Assert.AreEqual(LAT, actualLat);
        Assert.AreEqual(LON, actualLon);

    }
    ...
```

You can directly spot a few differences between the two code snippets:

- The unit test class in MSTest has to be decorated with TestClass.

- The constructor in MSTest would work, but the standard way of initializing is to decorate a method with TestInitialize.

- Both libraries use the Assert class name, but the method names in the class are different; for example, xUnit uses Equal and True, while MSTest uses AreEqual and IsTrue.

When doing multiple tests, xUnit and MSTest use different attributes. This code is in xUnit:

```
[Theory]
[InlineData("Freezing", -1)]
[InlineData("Scorching", 46)]
public async Task GetForecastWeatherAnalysis_
    Summary_MatchesTemp(string summary, double temp)
{
...
```

In MSTest, the equivalent code will look like this:

```
[DataTestMethod]
[DataRow("Freezing", -1)]
[DataRow("Scorching", 46)]
public async Task GetForecastWeatherAnalysis_
    Summary_MatchesTemp(string summary, double temp)
{
...
```

Here, you can notice two differences:

- `Theory` becomes `DataTestMethod`.

- `InlineData` becomes `DataRow`.

As you can see, there isn't much difference between the two libraries. Also, executing the test running, running the Test Explorer and other test activities other than the code stay the same.

NUnit

NUnit used to be the dominant library in the first decade of the two-thousands; it is still in use today, with xUnit becoming more prevalent.

To add an NUnit project to your solution, you can do it via the UI:

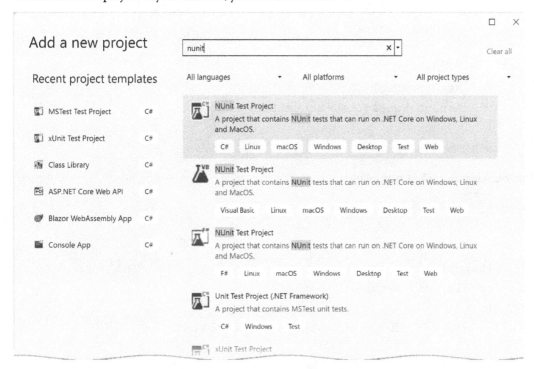

Figure A1.2 – Adding NUnit via the UI

Just like MSTest, NUnit has two copies of the .NET version. The bottom one is for the classical .NET Framework, and the top one is what we use with .NET Core.

You can add NUnit via the `dotnet` CLI:

```
dotnet new nunit
```

NUnit and xUnit have similar syntax, so let me show you the code in NUnit and its equivalent in MSTest. I'll start with xUnit:

```
public class WeatherAnalysisServiceTests
{
    ...

    public WeatherAnalysisServiceTests()
    {
        _sut = new (_openWeatherServiceMock);
    }


    [Fact]
    public async Task GetForecastWeatherAnalysis_
        LatAndLonPassed_ReceivedByOpenWeatherAccurately()

        ...
        // Assert
        Assert.Equal(LAT, actualLat);
        Assert.Equal(LON, actualLon);
    }
    ...
```

The equivalent in MSTest is as follows:

```
public class WeatherAnalysisServiceTests
{
    ...

    [Setup]
    public void Setup()
    {
        _sut = new(_openWeatherServiceMock);
    }


    [Test]
    public async Task GetForecastWeatherAnalysis_
      LatAndLonPassed_ReceivedByOpenWeatherAccurately()
    {
        ...
        // Assert
```

```
        Assert.That(actualLat, Is.EqualTo(LAT));
        Assert.That(actualLon, Is.EqualTo(LON));
    }
    ...
```

You can directly spot a few differences between the two code snippets:

- The constructor in NUnit would work, but the standard way of initializing is to decorate a method with `Setup`.
- Both libraries use the `Assert` class name, but the method names in the class are different; for example, xUnit uses `Equal`, while NUnit uses `AreEqual`.
- The style of NUnit uses a fluent interface design, and the recommended approach to test equality is to use `That` and `Is.EqualTo`.

When doing multiple tests, xUnit and NUnit use different class names. This code is in xUnit:

```
[Theory]
[InlineData("Freezing", -1)]
[InlineData("Scorching", 46)]
public async Task GetForecastWeatherAnalysis_
    Summary_MatchesTemp(string summary, double temp)
{
...
```

In NUnit, the equivalent code will look like this:

```
[Theory]
[TestCase("Freezing", -1)]
[TestCase("Scorching", 46)]
public async Task GetForecastWeatherAnalysis_
    Summary_MatchesTemp(string summary, double temp)
{
...
```

Here, you can notice that `InlineData` becomes `TestCase`. Besides this, there isn't much difference between the two libraries, and their template is included in the default installation of VS 2022.

These three libraries are interchangeable, and the syntax changes are minimal. Once you are used to one, it will take minimal time to switch to the other.

Mocking libraries

There is no shortage of mocking libraries in .NET; however, the top two used libraries are NSubstitute and Moq. We have covered plenty of examples of NSubstitute in this book, so let's see how Moq works.

Moq

Moq has the same role and same functionality, more or less, as NSubstitute. Given that the book was using NSubstitute, the fastest way to introduce Moq is to compare the two libraries. Let's start with a snippet from NSubstitute:

```
private IOpenWeatherService _openWeatherServiceMock =
    Substitute.For<IOpenWeatherService>();
private WeatherAnalysisService _sut;
private const decimal LAT = 2.2m;
private const decimal LON = 1.1m;
public WeatherAnalysisServiceTests()
{
    _sut = new (_openWeatherServiceMock);
}
[Fact]
public async Task GetForecastWeatherAnalysis_
    LatAndLonPassed_ReceivedByOpenWeatherAccurately()
{
    // Arrange
    decimal actualLat = 0;
    decimal actualLon = 0;
    _openWeatherServiceMock.OneCallAsync(
        Arg.Do<decimal>(x => actualLat = x),
        Arg.Do<decimal>(x => actualLon = x),
        Arg.Any<IEnumerable<Excludes>>(),
        Arg.Any<Units>())
        .Returns(Task.FromResult(GetSample(_defaultTemps)));

    // Act
    await _sut.GetForecastWeatherAnalysis(LAT, LON);

    // Assert
```

```
        Assert.Equal(LAT, actualLat);
        Assert.Equal(LON, actualLon);
}
```

This snippet instantiates and creates a mock object from `IOpenWeatherService` and spies on the `lat` and `lon` input parameters of `OneCallAsync`. The idea is to ensure the two parameters passed to `GetForecastWeatherAnalysis` are passed without modification to the `OneCallAsync` method.

Let's look at the same code using Moq:

```
private IOpenWeatherService _openWeatherServiceMock =
    Mock.Of<IOpenWeatherService>();
private WeatherAnalysisService _sut;
private const decimal LAT = 2.2m;
private const decimal LON = 1.1m;
public WeatherAnalysisServiceTests()
{
    _sut = new (_openWeatherServiceMock);
}
[Fact]
public async Task GetForecastWeatherAnalysis_
    LatAndLonPassed_ReceivedByOpenWeatherAccurately()
{
    // Arrange
    decimal actualLat = 0;
    decimal actualLon = 0;
    Mock.Get(_openWeatherServiceMock)
        .Setup(x => x.OneCallAsync(It.IsAny<decimal>(),
        It.IsAny<decimal>(),
        It.IsAny<IEnumerable<Excludes>>(),
        It.IsAny<Units>()))
        .Callback<decimal, decimal,
        IEnumerable<Excludes>, Units>((lat, lon, _, _) => {
            actualLat = lat; actualLon = lon; })
        .Returns(Task.FromResult(GetSample(_defaultTemps)));
```

```
    // Act
    await _sut.GetForecastWeatherAnalysis(LAT, LON);

    // Assert
    Assert.Equal(LAT, actualLat);
    Assert.Equal(LON, actualLon);
}
```

This Moq code doesn't look much different from its NSubstitute rival. Let's analyze the differences:

- NSubstitute instantiates a mock object using the `Substitute.For` method, whereas Moq does it using `Mock.Of`.

- NSubstitute uses extension methods such as `Returns` to configure the mock object, whereas Moq doesn't use extensions.

- NSubstitute uses `Args.Any` to pass parameters, whereas Moq uses `It.IsAny`.

In general, while Moq favors syntax with lambda expressions, NSubstitute takes another route and uses extension methods. NSubstitute tries to make the code look as natural as possible and get out of the way by having less syntax, while Moq relies on the power of lambdas.

> **Important note**
> Moq has another way of creating a mock. I opted to show the modern version.

In my opinion, using one library or another is a matter of style and syntax preference.

Unit testing helper libraries

I have seen developers adding these two libraries to their unit tests to enhance the syntax and readability: **Fluent Assertions** and **AutoFixture**.

Fluent Assertions

Fluent implementation, also known as a fluent interface, is trying to make the code read like an English sentence. Take this example:

```
Is.Equal.To(...);
```

Some developers like to have the tests written in this way as it supports a more natural way of reading a test. Some like it for their own reasons.

`FluentAssertions` is a popular library that integrates with all popular test frameworks among MSTest, Nunit, and xUnit to enable fluent interfaces. You can add it to your unit test project via NuGet under the name `FluentAssertions`.

Let's see how our code will be without and with the library:

```
// Without
Assert.Equal(LAT, actualLat);
// With
actualLat.Should().Be(LAT);
```

But the previous snippet doesn't show the true power of the library, so let's do some other examples:

```
// Arrange
string actual = "Hi Madam, I am Adam";
// Assert
actual.Should().StartWith("Hi")
    .And.EndWith("Adam")
    .And.Contain("Madam")
    .And.HaveLength(19);
```

The previous snippet is an example of a fluent syntax, and the code is self-explanatory. To test this code, you will require a few lines of the standard `Assert` syntax.

Here is another example:

```
// Arrange
var integers = new int[] { 1, 2, 3 };

// Assert
integers.Should().OnlyContain(x => x >= 0);
integers.Should().HaveCount(10,
  "The set does not contain the right number of elements");
```

The previous code is also self-explanatory.

> **Important Note**
> While these code snippets show the power of `FluentAssertions`, asserting too many unrelated elements in a unit test is not recommended. These examples are for illustration only and do not focus on best unit testing practices.

These two code snippets are enough to show why some developers are fond of this syntax style. Now that you know about it, the choice of using such syntax is yours.

AutoFixture

Sometimes, you have to generate data to populate an object. The object may be directly related to your unit test. Or maybe you just want to populate it for the rest of the unit to execute, but it is not the subject of the test. This is when **AutoFixture** comes to the rescue.

You can write the tedious code to generate an object, or you can use `AutoFixture`. Let's illustrate this with an example. Consider the following `record` class:

```
public record OneCallResponse
{
    public double Lat { get; set; }
    public double Lon { get; set; }

    ...

    public Daily[] Daily { get; set; }
}
public record Daily
{
    public DateTime Dt { get; set; }
    public Temp Temp { get; set; }

    ...

}
// More classes
```

Populating this in the `Arrange` part of your unit test will increase the size of your unit test and distract the test from its real intention.

AutoFixture can create an instance of this class using the least amount of code:

```
var oneCallResponse = _fixture.Create<OneCallResponse>();
```

This will create an object of this class and populate it with random values. These are some of the values:

```
{OneCallResponse { Lat = 186, Lon = 231, Timezone =
Timezone9d27503a-a90d-40a6-a9ac-99873284edef, TimezoneOffset =
177, Daily = Uqs.WeatherForecaster.Daily[] }}
    Daily: {Uqs.WeatherForecaster.Daily[3]}
    EqualityContract: {Name = "OneCallResponse" FullName =
        "Uqs.WeatherForecaster.OneCallResponse"}
```

```
Lat: 186
Lon: 231
Timezone: "Timezone9d27503a-a90d-40a6-a9ac-99873284edef"
TimezoneOffset: 177
```

The previous output is the first level of the `OneCallResponse` class, but all the succeeding levels are also populated.

But what if you want fine control of the generated data? Let's say we want to generate the class, but with the `Daily` property having an array length of 8 rather than a random size:

```
var oneCallResponse = _fixture.Build<OneCallResponse>()
    .With(x => x.Daily,_fixture.CreateMany<Daily>(8).ToArray())
    .Create();
```

This will generate everything randomly, but the `Daily` property will have eight array elements with random values.

This library has plenty of methods and customizations; this section only scratches the surface.

This appendix is a brief pointer to several libraries used for, or in conjunction with, unit testing. The intention here is to tell you these libraries exist and intrigue you to dig further if the need arises.

Further reading

To learn more about the topics discussed in the chapter, you can refer to the following links:

- *xUnit*: `https://xunit.net`
- *MSTest*: `https://docs.microsoft.com/en-us/dotnet/core/testing/unit-testing-with-mstest`
- *NUnit*: `https://nunit.org`
- *Moq*: `https://github.com/moq/moq4`
- *Fluent Assertions*: `https://fluentassertions.com`
- *AutoFixture*: `https://github.com/AutoFixture`

Appendix 2:
Advanced Mocking Scenarios

This book has numerous examples of straightforward mocking scenarios. And the good news is, in a clean code environment, implementing most of those mocking requirements will be easy.

However, there are times when you have to *innovate* a bit to be able to mock your desired class. I did not want to end this book without presenting you with a scenario, so here you go.

In this appendix, we will experience how to combine a fake with a mock to deal with a .NET class called `HttpMessageHandler`. By the end of this appendix, you will be familiar with more NSubstitute functionalities and ready to tackle more advanced mocking cases.

Technical requirements

The code for this chapter can be found at the following GitHub repository:

```
https://github.com/PacktPublishing/Pragmatic-Test-Driven-Development-
in-C-Sharp-and-.NET/tree/main/appendix2
```

Writing an OpenWeather client library

We have used the OpenWeather service in this book frequently, so I will give you a quick reminder – OpenWeather provides a set of RESTful APIs to bring you the weather and the forecast.

To consume the service from a C# application, it is good to have a library that will translate RESTful API calls to C# and shield the developer from the little details of HTTP. We call this sort of library a **RESTful client library** or sometimes a **software development kit** (**SDK**).

We will build an SDK for this service using TDD (of course!) and, while doing so, we will encounter more advanced mocking requirements.

The One Call API

OpenWeather has an API called **One Call** that will get you today's weather information and the forecast for the next few days. The best way to illustrate how this works is by using an example that gets the weather and the forecast for Greenwich, London.

First, issue a RESTful GET request. You can use your browser for this:

```
https://api.openweathermap.org/data/2.5/onecall?
    lat=51.4810&lon=0.0052&appid=[yourapikey]
```

Notice that the first two query string parameters are the latitude and the longitude of Greenwich, and the last one is your API key (which is omitted here). You will get a response similar to this:

```
{
    "lat":51.481,
    "lon":0.0052,
    "timezone":"Europe/London",
    "timezone_offset":3600,
    "current":{
        "dt":1660732533,
        "sunrise":1660711716,
        "sunset":1660763992,
        "temp":295.63,
        "feels_like":295.76,
        "pressure":1011,
        "humidity":70,
...
```

This is a very long JSON output; it is around 21,129 characters.

Creating the solution skeleton

We have created a library and tested it frequently, so we will need to do the same here:

1. Create a library project, call it Uqs.OpenWeather, and delete the sample class.
2. Create an xUnit project and call it Uqs.OpenWeather.Test.Unit.
3. Add a reference from the test project to the library.
4. Add NSubstitute from NuGet to the test project.
5. Rename the class and the filename in the unit test to ClientTests.cs.

Your VS solution will look like this:

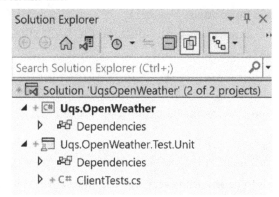

Figure A2.1 – Project skeleton's Solution Explorer

We are now ready to write the first unit test with TDD.

Starting the implementation with TDD

At this moment, you can open your `ClientTests.cs` and start your first test, which will drive the library's architecture.

We want to pass to a C# method, which we will call `OneCallAsync`, the two required parameters, `lat` and `lon`. This will then generate a URL with the right query string. So, our unit test class and the first unit test code will start to take shape, as shown in the following code base:

```
public class ClientTests
{
    private const string ONECALL_BASE_URL =
        "https://api.openweathermap.org/data/2.5/onecall";
    private const string FAKE_KEY = "thisisafakeapikey";
    private const decimal GREENWICH_LATITUDE = 51.4769m;
    private const decimal GREENWICH_LONGITUDE = 0.0005m;
    [Fact]
    public async Task
    OneCallAsync_LatAndLonPassed_UrlIsFormattedAsExpected()
    {
        // Arrange
        var httpClient = new HttpClient();
        var client = new Client(FAKE_KEY, httpClient);
```

```
         // Act
         var oneCallResponse = await
             client.OneCallAsync(GREENWICH_LATITUDE,
             GREENWICH_LONGITUDE);

         // Assert
         // will need access to the generated URL

     }
 }
```

Given that the API key needs to be sent with every API call, the API key is to be in the constructor and not part of the method parameters.

> **Important Note**
>
> Having the API key in the constructor will free the class's consumer of having to fetch the API key to pass it to the method call. Instead, it will become the responsibility of the dependency injection setup to fetch the key, which makes more sense.

We will definitely need the `HttpClient` class because your client will be using REST, and this is what you usually use with RESTful calls in .NET Core. However, when using this class, we may face the following challenges:

- `HttpClient` is a concrete class, and calling any method on it will lead to `HttpClient` issuing a call to the destination – this is the default behavior, but it can be tweaked.

- `HttpClient` doesn't give us access to the generated URL, which is what we want in the current test.

We need to figure out a way to intercept the call before `HttpClient` calls the destination (which is the actual third-party service) and get a hold of the generated URL for inspection. Of course, we also want to eliminate the outbound call as this is a unit test, and we don't want to call the third party for real.

`HttpClient` can be passed an instance of `HttpMessageHandler` in the constructor, and then from `HttpMessageHandler` we can get hold of the generated URL by spying on `HttpMessageHandler`. `SendAsync` and eliminate the real call. But `HttpMessageHandler` is an abstract class, so we cannot instantiate it; we need to inherit from it.

So, let's create a child class from `HttpMessageHandler` and call it `FakeHttpMessageHandler` in your unit test project, as follows:

```
public class FakeHttpMessageHandler : HttpMessageHandler
{
```

```
    private HttpResponseMessage _fakeHttpResponseMessage;
    public FakeHttpMessageHandler(
        HttpResponseMessage responseMessage)
    {
        _fakeHttpResponseMessage = responseMessage;
    }
    protected override Task<HttpResponseMessage>
        SendAsync(HttpRequestMessage request,
        CancellationToken cancellationToken)
    => SendSpyAsync(request, cancellationToken);

    public virtual Task<HttpResponseMessage>
        SendSpyAsync(HttpRequestMessage request,
        CancellationToken cancellationToken)
    => Task.FromResult(_fakeHttpResponseMessage);
}
```

We have created a fake class that will allow us to get access to `HttpRequestMessage`. Now, our Arrange will look like this:

```
// Arrange
var httpResponseMessage = new HttpResponseMessage()
{
    StatusCode = HttpStatusCode.OK,
    Content = new StringContent("{}")
};
var fakeHttpMessageHandler = Substitute.ForPartsOf
    <FakeHttpMessageHandler>(httpResponseMessage);
HttpRequestMessage? actualReqMessage = null;
fakeHttpMessageHandler.SendSpyAsync(
    Arg.Do<HttpRequestMessage>(x => actualReqMessage = x),
    Arg.Any<CancellationToken>())
    .Returns(Task.FromResult(httpResponseMessage));
var fakeHttpClient = new
  HttpClient(fakeHttpMessageHandler);
var client = new Client(FAKE_KEY, fakeHttpClient);
```

We first created a response message, so any method call will return this empty object. This object will contain the third-party response when we run the real code.

> **Important Note**
>
> We have created a fake for `HttpMessageHandler`; we could have mocked it as well. Both work, and it is based on preference of what is more readable. Here I feel that having a fake `HttpMessageHandler` is easier to read.
>
> Also note that the preceding implementation could be called a stub (rather than a fake) but I opted to refer to it as fake as it contains some real implementation. Sometimes there is a thin line between stubs and fakes.

Note that we used NSubstitute to create a mock of the fake. The reason for that is we want to have access to `HttpRequestMessage`, which contains our final URL.

We have used `Substitute.ForPartsOf` rather than `Substitute.For` for the first time in this book, because `For` is not meant for concrete classes; the code will compile, but you will get a runtime error.

> **Important Note**
>
> We have always used `Substitute.For<ISomeInterface>` and this is what you would do in 95% of the cases. We have not created an instance of a concrete class. For concrete classes, without interfaces, you would use `ForPartsOf<SomeClass>`.

Our `Assert` section becomes this:

```
string actualUrl = actualHttpRequestMessage!.RequestUri!
    .AbsoluteUri.ToString();
Assert.Contains(ONECALL_BASE_URL, actualUrl);
Assert.Contains($"lat={GREENWICH_LATITUDE}", actualUrl);
Assert.Contains($"lon={GREENWICH_LONGITUDE}", actualUrl);
```

Now, we are ready to write the production code.

Fail then pass

The code will fail to even compile as we have not created the production code, which will give us the TDD fail that we are looking for. Now, we shall do the minimal implementation that will pass the test:

```
public class Client
{
```

```
    ...
    public async Task<OneCallResponse> OneCallAsync(
        decimal latitude, decimal longitude)
    {
        const string ONECALL_URL_TEMPLATE = "/onecall";
        var uriBuilder = new UriBuilder(
          BASE_URL + ONECALL_URL_TEMPLATE);
        var query = HttpUtility.ParseQueryString("");
        query["lat"] = latitude.ToString();
        query["lon"] = longitude.ToString();
        query["appid"] = _apiKey;

        uriBuilder.Query = query.ToString();

        var _ = await _httpClient
            .GetStringAsync(uriBuilder.Uri.AbsoluteUri);

        return new OneCallResponse();
    }
}
```

Run your test again, and it will pass.

We've done a lot for this test, but other tests will flow easily as they will use the same fake that we've created. Let's do a recap of what we've done.

Recap

Here is a recap of all the important activities that we've done to make the first test pass:

- We wanted to write a test that checks whether the URL is formed right.

- We had to get into the internals of HttpClient to get the URL.

- HttpClient doesn't have the right methods to spy on the generated URL.

- We created a fake FakeHttpMessageHandler that inherited HttpMessageHandler and passed it to HttpClient so we can reach the HttpClient internals.

- We mocked our fake FakeHttpMessageHandler and spied on the URL.

- We utilized a less-used method of NSubstitute for creating a mock, `Substitute.ForPartsOf`, which allowed us to mock a concrete class.

- We followed the standard TDD route of fail and pass to implement our production code.

I hope this made the activities clearer. You have the full source code to inspect if this is not the case.

You will encounter similar advanced mocking scenarios in the future, so how do you attack them?

Investigating complex mocking scenarios

Like everything in a developer's life, you will be able to find someone else who has encountered a similar mocking scenario to the one you are facing. Searching online for `access url HttpClient NSubstitute` would have given you the clues you needed to cut through the problem quickly.

The good news is that most of the sophisticated mocking problems have already been sorted out, and the solutions have been published (thanks to all developers' hard work). You just need to understand the concept and incorporate it into your solution.

In this appendix, we went through a more advanced but less frequent mocking scenario. It requires more fiddling and extra effort, but with experience in mocking, you will become familiar with these scenarios, and you will cut through them in no time.

Further reading

To learn more about the topics discussed in the chapter, you can refer to *OpenWeather's* official website: `https://openweathermap.org`

Index

Packt.com

Subscribe to our online digital library for full access to over 7,000 books and videos, as well as industry leading tools to help you plan your personal development and advance your career. For more information, please visit our website.

Why subscribe?

- Spend less time learning and more time coding with practical eBooks and Videos from over 4,000 industry professionals

- Improve your learning with Skill Plans built especially for you

- Get a free eBook or video every month

- Fully searchable for easy access to vital information

- Copy and paste, print, and bookmark content

Did you know that Packt offers eBook versions of every book published, with PDF and ePub files available? You can upgrade to the eBook version at packt.com and as a print book customer, you are entitled to a discount on the eBook copy. Get in touch with us at customercare@packtpub.com for more details.

At www.packt.com, you can also read a collection of free technical articles, sign up for a range of free newsletters, and receive exclusive discounts and offers on Packt books and eBooks.

Other Books You May Enjoy

If you enjoyed this book, you may be interested in these other books by Packt:

Enterprise Application Development with C# 10 and .NET 6 - Second Edition

Ravindra Akella, Arun Kumar Tamirisa, Suneel Kumar Kunani and Bhupesh Guptha Muthiyalu

ISBN: 9781803232973

- Design enterprise apps by making the most of the latest features of .NET 6
- Discover different layers of an app, such as the data layer, API layer, and web layer
- Explore end-to-end architecture by implementing an enterprise web app using .NET and C# 10 and deploying it on Azure
- Focus on the core concepts of web application development and implement them in .NET 6
- Integrate the new .NET 6 health and performance check APIs into your app
- Explore MAUI and build an application targeting multiple platforms - Android, iOS, and Windows

High-Performance Programming in C# and .NET

Jason Alls

ISBN: 9781800564718

- Use correct types and collections to enhance application performance
- Profile, benchmark, and identify performance issues with the codebase
- Explore how to best perform queries on LINQ to improve an application's performance
- Effectively utilize a number of CPUs and cores through asynchronous programming
- Build responsive user interfaces with WinForms, WPF, MAUI, and WinUI
- Benchmark ADO.NET, Entity Framework Core, and Dapper for data access
- Implement CQRS and event sourcing and build and deploy microservices

Parallel Programming and Concurrency with C# 10 and .NET 6

Alvin Ashcraft

ISBN: 9781803243672

- Prevent deadlocks and race conditions with managed threading
- Update Windows app UIs without causing exceptions
- Explore best practices for introducing asynchronous constructs to existing code
- Avoid pitfalls when introducing parallelism to your code
- Implement the producer-consumer pattern with Dataflow blocks
- Enforce data sorting when processing data in parallel and safely merge data from multiple sources
- Use concurrent collections that help synchronize data across threads
- Debug an everyday parallel app with the Parallel Stacks and Parallel Tasks windows

Packt is searching for authors like you

If you're interested in becoming an author for Packt, please visit `authors.packtpub.com` and apply today. We have worked with thousands of developers and tech professionals, just like you, to help them share their insight with the global tech community. You can make a general application, apply for a specific hot topic that we are recruiting an author for, or submit your own idea.

Share Your Thoughts

Now you've finished *Pragmatic Test-Driven Development in C# and .NET*, we'd love to hear your thoughts! Scan the QR code below to go straight to the Amazon review page for this book and share your feedback or leave a review on the site that you purchased it from.

`https://packt.link/r/1803230193`

Your review is important to us and the tech community and will help us make sure we're delivering excellent quality content.